MOLECULAR
ELECTRONIC
STRUCTURES
AN INTRODUCTION

MOLECULAR ELECTRONIC STRUCTURES
AN INTRODUCTION

CARL J. BALLHAUSEN   *Københavens Universitet*

HARRY B. GRAY   *California Institute of Technology*

1980
THE BENJAMIN/CUMMINGS PUBLISHING COMPANY, INC.
ADVANCED BOOK PROGRAM
Reading, Massachusetts

London • Amsterdam • Don Mills, Ontario • Sydney • Tokyo

Carl J. Ballhausen and Harry B. Gray
Molecular Electronic Structures: An Introduction, 1980

Harry B. Gray is co-author of

Ligand Substitution Processes

by Cooper H. Langford and Harry B. Gray

1966 (Second printing, with revisions, 1974), W. A. Benjamin, Inc., Advanced Book Program, Reading, Massachusetts.

Library of Congress Cataloging in Publication Data

Ballhausen, Carl Johan, 1926–
    Molecular electronic structures.

    Bibliography: p.
    Includes index.
    1. Molecular bonds.   2. Molecular structure.
I. Gray, Harry B., joint author.   II. Title.
QD461.B23   1980   541.2'24   80-17429
ISBN 0-8053-0452-5

ABCDEFGHIJK-AL-89876543210

Manufactured in the United States of America

CONTENTS

PREFACE

The present book is an introduction to molecular electronic structural theory. It is aimed at students who have reasonable familiarity with differential and integral calculus and are beginning a study of the physical description of chemical systems. We have decided to concentrate on the description of ground state electronic structures, or, in other words, the principles of chemical bonding in molecules. In this important respect the present volume differs from our earlier book "Molecular Orbital Theory" (Benjamin, 1964), which included a strong emphasis on the description of electronic excited states.

In our treatment of molecular wave functions we make use of "symmetry operators", the latter being operators that leave the Hamiltonian unchanged. By using such symmetry operators, it is possible to characterize the electronic structures of molecules. In our opinion, this approach provides good preparation for later studies that may be undertaken in which formal group theory is employed.

The heart of the book is Chapter 4, where we discuss in some detail the bonding in several selected molecules. Examples from both organic and inorganic chemistry are included in an attempt to make the coverage as general as possible. Our objective here is to provide an introduction to molecular bonding that will serve as a foundation for more advanced study of electronic structures.

Suggested reading and problems are included in each chapter. Some of the problems are challenging, but working them will give the student a much better feeling for the principles involved. The suggested reading is of two types, books (and reviews) and original papers. And we urge students to examine at least some of the older papers in the field, as much can be learned from them.

C. J. BALLHAUSEN
HARRY B. GRAY
Pasadena, California

ACKNOWLEDGMENT

Most of this book was written while C.J.B. was a Sherman Fairchild Distinguished Scholar at the California Institute of Technology. C.J.B. wishes to express his sincere thanks to President Marvin L. Goldberger of Caltech for the invitation to spend parts of 1979 and 1980 in Pasadena. Assistance at Caltech provided by Lea Sterret, Carol Cooper, Pat Bullard, and Jennifer Burkhart is much appreciated.

THE SCHRÖDINGER EQUATION

1-1 FORMALISM

One of the important problems confronting the chemist is that of understanding how atoms are "bonded" together to form molecules. Quantum theory has proved that the stability of molecules and ions is a consequence of the motion of electrons in the coulomb potential field stemming from the nuclei. The stable molecular system has an electronic ground state that is associated with a certain molecular shape. A molecule also has electronic excited states. These latter states may or may not have an energy minimum at the stable nuclear configuration of the ground state.

The behavior of electrons is governed by an equation formulated by E. Schrödinger and which bears his name. In a molecular description of nature, this equation occupies the same place as Newton's second law does in a macroscopic description. However, the presence of Planck's constant, h, in the Schrödinger equation shows that this equation cannot be derived from classical mechanics. This follows because all observable quantities that contain Planck's constant are discontinuous and, consequently, not classical. It can be shown, however, that the quantum description goes asymptotically into the classical laws of motion.

The Schrödinger equation for a single particle of mass m moving in a potential field $V(x,\text{y},z)$ is

$$\left[-\frac{\hbar^2}{2m} \nabla^2 + V(x,\text{y},z) \right] \psi(x,y,z) = W\psi(x,y,z). \quad (1\text{-}1)$$

W is the total (time-independent) energy of the particle,

$$\nabla^2 = \left(\frac{\partial^2}{\partial x^2} + \frac{\partial^2}{\partial y^2} + \frac{\partial^2}{\partial z^2} \right) \tag{1-2}$$

and $\hbar = h/2\pi$. The second-order differential equation (1-1) has many solutions, $\psi_j(x,y,z), j = 0,1,2,\ldots$. These are called the *wavefunctions* or *orbitals* for the particle. The functions $\psi_j(x,y,z)$ may be complex. They account for the behavior of the particle in the following sense. The *probability* of finding the particle described by $\psi_j(x,y,z)$ in space between x and $x + dx$, y and $y + dy$ and z and $z + dz$ is given by $|\psi_j(x,y,z)|^2 d\tau$, where $d\tau = dxdydz$. $|\psi_j(x,y,z)|^2$ is equal to $\psi_j(x,y,z) \cdot \psi_j^*(x,y,z)$, that is, the function multiplied by its complex conjugate.[1] This procedure ensures that $|\psi_j|^2$ is a real function.

We observe from the form of the Schrödinger equation (1-1) that provided $\psi(x,y,z)$ is a solution, $N\psi(x,y,z)$, where N is a constant, is also a solution. We can determine N by demanding that the total probability of finding the particle somewhere in the space it can occupy is unity; that is,

$$\iiint |\psi_j(x,y,z)|^2 \, dx \, dy \, dz = 1. \tag{1-3}$$

We say that we have *normalized* the wavefunction to one.

That the functions should be integrable imposes the following conditions on the solution of Eq. (1-1) in order that $\psi_j(x,y,z)$ may be acceptable wavefunctions for the particle: They must be finite, single-valued, and continuous at every point in space.

The total energy, W, of the particle may or may not assume discrete values. A discussion of the solutions to the Schrödinger equation that have a continuous range of energies is, however, outside the scope of this book. In all cases in which we shall be

---

[1]Given a complex function $\psi(x,y,z) = g(x,y,z) + if(x,y,z)$, $i = \sqrt{-1}$. The complex conjugate of any function is obtained by changing the sign of i, whenever it occurs, in this case $g(x,y,z) - if(x,y,z)$. Then $\psi(x,y,z) \cdot \psi^*(x,y,z) = g^2(x,y,z) + f^2(x,y,z)$, a real function.

interested, the Schrödinger equation (1-1) for a particle in a given potential field $V(x,y,z)$ yields a discrete set of energies W_j and a corresponding set of acceptable solutions $\psi_j(x,y,z)$.

For a certain energy W_j, it may happen that there is more than one acceptable solution to the Schrödinger equation. We call such an energy level *degenerate*, the degeneracy being equal to the number of linear independent wavefunctions corresponding to that energy. Linear independence of the wavefunctions is mathematically expressed by

$$\iiint \psi_j^*(x,y,z) \cdot \psi_k(x,y,z)\, dx\, dy\, dz = 0, \qquad j \neq k. \quad (1\text{-}4)$$

We say that ψ_j and ψ_k are *orthogonal* to each other.

Let us suppose we have solved the Schrödinger equation for a certain system and obtained a set of wavefunctions, $\psi_j(x,y,z)$, $j = 0,1,2,\ldots$. Since the physical state of the system is completely determined by the ψ's we have to be able to use these ψ's to calculate physical quantities pertinent to the system. In quantum mechanics we have that for every observable quantity a_j there exists a corresponding *operator*, $\hat{\alpha}$. We calculate a_j by operating on the wavefunction ψ_j with $\hat{\alpha}$:

$$\hat{\alpha}\psi_j = a_j\psi_j.$$

a_j is called the *eigenvalue* and ψ_j the *eigenfunction* of the operator $\hat{\alpha}$. In order to ensure that a_j is a real number $\hat{\alpha}$ must be a *Hermitian operator*, that is,

$$\int \omega_k^*\hat{\alpha}\psi_j\, d\tau = \int \psi_j\hat{\alpha}\psi_k^*\, d\tau.$$

Then, using normalized wavefunctions,

$$\int \psi_j^*\hat{\alpha}\psi_j\, d\tau = \int \psi_j^*a_j\psi_j\, d\tau = a_j \int \psi_j^*\psi_j\, d\tau = a_j,$$
$$\parallel$$
$$\int \psi_j\hat{\alpha}^*\psi_j^*\, d\tau = \int \psi_j(\hat{\alpha}\psi_j)^*\, d\tau = \int \psi_j a_j^*\psi_j^*\, d\tau = a_j^*$$

or $a_j = a_j^*$, that is, a_j must be real.

Further, we have that the eigenfunctions of a Hermitian operator form an orthogonal set of functions. In the case where we do not have any degeneracies, this is proven as follows. Let

$$\hat{\alpha}\psi_k = a_k \psi_k \quad \text{and} \quad \hat{\alpha}\psi_j = a_j\psi_j, \qquad a_k \neq a_j.$$

Then

$$\int \psi_j^* \hat{\alpha} \psi_k \, d\tau = \int \psi_k \hat{\alpha}^* \psi_j^* d\tau$$

$$\|\qquad\qquad\qquad\qquad \|$$

$$a_k \int \psi_j^* \psi_k \, d\tau \quad a_j^* \int \psi_k \psi_j^* d\tau$$

Since $a_j^* = a_j$, because $a3$ is a Hermitian operator, we have

$$(a_k - a_j) \int \psi_j^* \psi_k \, d\tau = 0$$

or, since $a_k \neq a_j$,

$$\int \psi_j^* \psi_k \, d\tau = 0.$$

Introducing the so-called Hamiltonian operator,

$$\mathscr{H} = - \frac{\hbar^2}{2m} \nabla^2 + V(x,y,z) \qquad (1\text{-}5)$$

the Schrödinger equation (1-1) can be written in the following condensed form

$$\mathscr{H}\psi_j = W_j\psi_j. \qquad (1\text{-}6)$$

$\mathscr{H}$ is a Hermitian operator. The $\psi_j(x,y,z), j = 0,1,2,\dots$ solutions to the same Schrödinger equation therefore constitute an orthogonal set. We can write

$$\int \psi_j^* \psi_k \, d\tau = \delta_{jk}$$

using the so-called Kronecker delta:

$$\delta_{j,k} = \begin{cases} 1 & \text{for } j = k, \\ 0 & \text{for } j \neq k. \end{cases}$$

By multiplying "from the left" in (1-6) with ψ^* and integrating, we obtain

$$\frac{\int \psi_j^* \mathscr{H} \psi_j \, d\tau}{\int \psi_j^* \psi_j \, d\tau} = W_j, \qquad j = 0, 1, 2, \ldots . \quad (1\text{-}7)$$

Provided ψ_j is normalized to one, we find

$$\int \psi_j^* \mathscr{H} \psi_j \, d\tau = W_j. \quad (1\text{-}8)$$

Obviously, $\mathscr{H}$ is the operator that gives the energy of the system. Classically, the kinetic energy of a particle is given by $p^2/2m$, where p is the linear momentum ($p = mv$). In quantum mechanics we can indentify $\hat{p}_x$ with the operator $(\hbar/i)(\partial/\partial x)$,

$$\hat{p}_y = \frac{\hbar}{i} \frac{\partial}{\partial y} \quad \text{and} \quad \hat{p}_z = \frac{\hbar}{i} \frac{\partial}{\partial z}, \quad \text{where } i = \sqrt{-1}.$$

With $p^2 = p_x^2 + p_y^2 + p_z^2$, the two operator terms in (1-5) represent, therefore, the kinetic energy and the potential energy of the particle, respectively.

In case a wavefunction is not an eigenfunction of an operator $\hat{\alpha}$, we get the *expectation value*, $\langle a_j \rangle$, by a calculation using the following expression:

$$\langle a_j \rangle = \frac{\int \psi_j^* \hat{\alpha} \psi_j \, d\tau}{\int \psi_j^* \psi_j \, d\tau}$$

which, in case ψ_j is normalized, becomes

$$\langle a_j \rangle = \int \psi_j^* \hat{a} \psi_j \, d\tau. \qquad (1\text{-}9)$$

Let us now assume that we are unable to solve the Schrödinger equation for a certain system, but that we can make a more or less intelligent guess at a solution, $\phi(x,y,z)$. We then take the function $\phi(x,y,z)$ and calculate

$$\langle W \rangle = \frac{\displaystyle\int \phi^* \mathscr{H} \phi \, d\tau}{\displaystyle\int |\phi|^2 \, d\tau}. \qquad (1\text{-}10)$$

If W_0 in Eq. 1-7 is the lowest eigenvalue of the $\mathscr{H}$ operator, then we shall show that for any guessed function, $\phi(x,y,z)$, we will always have $\langle W \rangle \geq W_0$. The better our trial function $\phi(x,y,z)$, the closer the $\langle W \rangle$ will be to W_0. This is the *variational principle*. We will prove it as follows:

Suppose we could solve the Schrödinger equation for a specific case. We would then get a set of wavefunctions and eigenvalues (ψ_0, W_0), (ψ_1, W_1), (ψ_2, W_2), etc. We exclude the possibility of degeneracy and take $W_0 < W_1 < W_2 \ldots$. Without loss of generality we can take the set ψ_j to be real and *expand* $\phi(x,y,z)$ on the orthogonal and normalized set $\psi_j(x,y,z)$

$$\phi = \sum_{j=0}^{\infty} a_j \psi_j. \qquad (1\text{-}11)$$

Then, substituting (1-11) into (1-10), we find

$$\langle W \rangle = \frac{\displaystyle\int \left(\sum_{j=0}^{\infty} a_j \psi_j \right) \mathscr{H} \left(\sum_{j=0}^{\infty} a_j \psi_j \right) d\tau}{\displaystyle\int \left(\sum_{j=0}^{\infty} a_j \psi_j \right) \left(\sum_{j=0}^{\infty} a_j \psi_j \right) d\tau},$$

$$\langle W \rangle = \frac{\int \left(\sum\limits_{j=0}^{\infty} a_j \psi_j \right) \left(\sum\limits_{j=0}^{\infty} W_j a_j \psi_j \right) d\tau}{a_0^2 + a_1^2 + a_2^2 + \cdots} ,$$

$$\langle W \rangle = \frac{W_0 a_0^2 + W_1 a_1^2 + W_2 a_2^2 + \cdots}{a_0^2 + a_1^2 + a_2^2 + \cdots} .$$

Since $W_0 < W_1 < W_2 \cdots$, we have

$$\langle W \rangle \geq \frac{W_0 (a_0^2 + a_1^2 + a_2^2 + \cdots)}{(a_0^2 + a_1^2 + a_2^2 + \cdots)} = W_0.$$

In practice $\phi(x,y,z)$ is taken as a function $\phi(x,y,z,c_i)$ which contains certain parameters, c_i. One therefore calculates $\langle W \rangle$ as a function of these parameters. Afterwards one minimizes $\langle W \rangle$ with respect to the parameters, thereby ensuring that the lowest value of $\langle W \rangle$ is obtained. This minimized $\langle W \rangle$ is then the best approximation we can obtain for W_0 with the given variational function, and the function $\phi(x,y,z,\tilde{c}_i)$ gives the best approximation to the "true" wavefunction ψ_0.

1-2 THE PARTICLE IN A BOX

As a simple example of the application of the Schrödinger equation we shall treat a particle of mass m in a one-dimensional box of length L. The potential energy V is constant inside the box, and can be put equal to zero. Outside the box we have $V = \infty$. The Hamiltonian operator is then

$$\mathscr{H} = -\left(\frac{\hbar^2}{2m} \right) \frac{d^2}{dx^2} , \qquad 0 \leq x \leq L. \qquad (1\text{-}12)$$

ψ is of course different from zero inside the box, but is zero outside, since the particle cannot move in that region and consequently cannot be found there. The continuity of $\psi(x)$ then demands that $\psi(0) = \psi(L) = 0$.

The Schrödinger equation is

$$-\frac{\hbar^2}{2m}\frac{d^2\psi(x)}{dx^2} = W\psi(x), \qquad 0 \le x \le L. \qquad (1\text{-}13)$$

The equation is a second-order differential equation and we solve it by guessing at a function with so many parameters that we can fit it to the boundary conditions.

$$\psi(x) = A \sin \alpha x + B \cos \beta x.$$

By using our conditions $\psi(0) = \psi(L) = 0$, we get

$$\psi(0) = A \sin(\alpha \cdot 0) + B \cos(\beta \cdot 0) = 0.$$

The above equation has the solution $B = 0$. Further,

$$\psi(L) = 0 = A \sin \alpha L.$$

This means

$$\alpha L = n\pi \qquad n = 0,1,2,\ldots$$

giving

$$\psi(x) = A \sin \frac{n\pi}{L} x. \qquad (1\text{-}14)$$

We have, then,

$$-\frac{d^2\psi}{dx^2} = A \frac{n^2\pi^2}{L^2} \sin \frac{n\pi}{L} x = \frac{n^2\pi^2}{L^2} \psi(x).$$

Inserting in (1-13) gives

$$W = \frac{h^2 n^2}{8mL^2} . \qquad (1\text{-}15)$$

$|\psi|^2 dx$ is the probability of finding the particle between x and dx, and since we know that the particle is in the box the normalization condition demands

$$\int_0^L |\psi|^2 \, d\tau = 1. \qquad (1\text{-}16)$$

Hence

$$A^2 \int_0^L \sin^2 \frac{n\pi}{L} x \, dx = 1$$

or

$$A = \left(\frac{2}{L} \right)^{1/2}.$$

The normalized wave functions and energies for this system are then

$$\psi_n(x) = \left(\frac{2}{L} \right)^{1/2} \sin \frac{n\pi}{L} x; \qquad W_n = \frac{h^2 n^2}{8mL^2} \qquad (1\text{-}17)$$

$$n = 1, 2, \ldots$$

since $n = 0$ is without physical meaning. The wavefunctions $\psi(x)$ are shown in Figure 1-1: Notice that the more nodes an orbital possesses, the higher its energy. This is a quite general feature.

The expectation value of the operators $\hat{x}$ and the linear momentum $\hat{p}_x = (\hbar/i)(d/dx)$ are now calculated for a particle in the box. We get

$$\langle x \rangle = \int_0^L \frac{2}{L} x \sin^2 \left(\frac{n\pi}{L} x \right) dx = \frac{L}{2},$$

$$\langle p_x \rangle = \int_0^L \frac{2}{L} \sin \left(\frac{n\pi}{L} x \right) \left(\frac{\hbar}{i} \frac{d}{dx} \right) \sin \left(\frac{n\pi}{L} x \right) dx = 0.$$

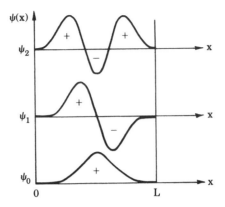

Figure 1-1 Wavefunctions for a particle in a one-dimensional box.

We have also

$$\langle x^2 \rangle = \int_0^L \frac{2}{L} x^2 \sin^2\left(\frac{n\pi}{L}x\right) \, dx = L^2\left(\frac{1}{3} - \frac{1}{2n^2\pi^2}\right),$$

$$\langle p_x^2 \rangle = \int_0^L \frac{2}{L} \sin\left(\frac{n\pi}{L}x\right)\left(-\hbar^2\frac{d^2}{dx^2}\right)\sin\left(\frac{n\pi}{L}x\right) \, dx$$

$$= \frac{\hbar^2\pi^2n^2}{L^2}.$$

With

$$\delta x \cdot \delta p_x = \sqrt{\langle x^2 \rangle - \langle x \rangle^2} \cdot \sqrt{\langle p_x^2 \rangle - \langle p_x \rangle^2}$$

we find

$$\delta x \cdot \delta p_x = \hbar \sqrt{\frac{\pi^2n^2}{12} - \frac{1}{2}}.$$

For the ground state $n = 1$ and $\delta x \delta p_x = \hbar \cdot 0.568$. This result can be compared with the *Heisenberg uncertainty principle*, which states that in general for any system $\delta x \delta p_x \geq \hbar /2$.

We shall now consider a particle in a three-dimensional rectangular box with the sides L_x, L_y, L_z. The Schrödinger equation is here

$$-\frac{\hbar^2}{2m} \left[\frac{\partial^2}{\partial x^2} + \frac{\partial^2}{\partial y^2} + \frac{\partial^2}{\partial z^2} \right] \psi(x,y,z) = W\psi(x,y,z).$$

$$(1-18)$$

The potential energy V is zero inside the box but infinite outside the box. As before we must therefore have $\psi(x,y,z) = 0$ on the walls of the box.

In order to solve the Schrödinger equation, we try a solution of the form

$$\psi(x,y,z) = X(x)Y(y)Z(z) \qquad (1-19)$$

and substitute

$$-\frac{\hbar^2}{2m} \left[Y(y)Z(z) \, \frac{d^2X(x)}{dx^2} + X(x)Z(z) \, \frac{d^2Y(y)}{dy^2} \right.$$

$$\left. + X(x)Y(y) \, \frac{d^2Z(z)}{dz^2} \right] = X(x)Y(y)Z(z)W. \quad (1-20)$$

We divide through with $X(x)Y(y)Z(z)$ and obtain

$$-\frac{\hbar^2}{2m} \left[\frac{1}{X(x)} \frac{d^2X(x)}{dx^2} + \frac{1}{Y(y)} \frac{d^2Y(y)}{dy^2} + \right.$$

$$\left. + \frac{1}{Z(z)} \frac{d^2Z(z)}{dz^2} \right] = W. \qquad (1-21)$$

We split the constant W into three terms

$$W = W_x + W_y + W_z \qquad (1-22)$$

and since the three coordinates x,y,z are completely independent of each other, we must have

$$-\frac{\hbar^2}{2m}\frac{X''}{X} = W_x, \qquad (1\text{-}23)$$

$$-\frac{\hbar^2}{2m}\frac{Y''}{Y} = W_y, \qquad (1\text{-}24)$$

$$-\frac{\hbar^2}{2m}\frac{Z''}{Z} = W_z. \qquad (1\text{-}25)$$

Each of these three equations can be solved as the one-dimensional case, and we obtain

$$W = W_x + W_y + W_z = \frac{h^2}{8m}\left[\frac{n_x^2}{L_y^2} + \frac{n_y^2}{L_y^2} + \frac{n_z^2}{L_z^2}\right].$$

$$(1\text{-}26)$$

Notice that we now have three quantum numbers in the expression for the energy, one for each of the three dimensions. The normalized wavefunction is

$$\psi(x,y,z) =$$

$$\left(\frac{8}{L_x L_y L_z}\right)^{1/2}\left(\sin\frac{\pi n_x}{L_x}x\right)\left(\sin\frac{\pi n_y}{L_y}y\right)\left(\sin\frac{\pi n_z}{L_z}z.\right)$$

$$(1\text{-}27)$$

In the case one side of the box, e.g., L_x, is much larger than L_y and L_z, as shown in Figure 1-2, we see that the energy levels written down from the lowest one and up are

$$(n_x, n_y, n_z) = (1, 1, 1)$$
$$(2, 1, 1)$$
$$(3, 1, 1)$$
$$(4, 1, 1)$$
$$\cdot$$
$$\cdot$$
$$\cdot$$

It is seen that we have many energy states before n_y and n_z move up above 1. Notice also that we can have a number of different wavefunctions that have the same energy. In a cubic box we will have three different wavefunctions with the energy $6h^2/8mL^2$, six wavefunctions with energy $14h^2/8mL^2$, and so forth. This is then an example of degeneracy.

Had we not been able to solve the Schrödinger equation for a particle in a one-dimensional box, we could in accordance with the variational principle have guessed at a function and then calculated the ground state energy. A suitable function, being zero at the walls $x = 0$ and $x = L$ is

$$\psi(x) = N(x-L)x. \qquad (1\text{-}28)$$

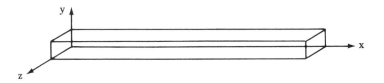

Figure 1-2 Rectangular box with $L_x \gg L_y \approx L_z$.

We have then:

$$\langle W \rangle = \frac{N^2 \left(-\dfrac{\hbar^2}{2m} \right) \displaystyle\int_0^L (x-L)x \; \dfrac{d^2}{dx^2}\Big((x-L)x\Big)dx}{N^2 \displaystyle\int_0^L (x-L)^2 x^2 \, dx} \tag{1-29}$$

$$= -\frac{\hbar^2}{2m} \frac{\displaystyle\int_0^L (x-L)x \cdot 2 \, dx}{\displaystyle\int_0^L (x-L)^2 x^2 \, dx}$$

$$= -\frac{\hbar^2}{m} \frac{\displaystyle\int_0^L x^2 \, dx - L \int_0^L x \, dx}{\displaystyle\int_0^L x^4 \, dx + L^2 \int_0^L x^2 \, dx - 2L \int_0^L x^3 \, dx}$$

$$= -\frac{\hbar^2}{mL^2}(-5) = 1.013 \; \frac{h^2}{8mL^2} \; .$$

Notice that the energy for the lowest state of the particle, estimated by the use of an approximate wavefunction, is only 1.3% higher than the true value.

1-3 ATOMIC ORBITALS

We shall now examine the solutions to the Schrödinger equation for an electron moving in a spherically symmetrical coulomb potential field such as occurs around an atomic nucleus. In this case it is convenient to use a system of polar coordinates (Figure 1-3).

The following conversions are easily obtained for the coordinates.

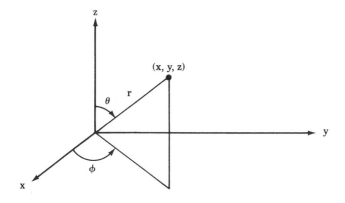

Figure 1-3 Polar and rectangular coordinates.

$$x = r \sin \theta \cos \phi, \tag{1-30}$$

$$y = r \sin \theta \sin \phi, \tag{1-31}$$

$$z = r \cos \theta. \tag{1-32}$$

More difficulty is encountered with the conversion of the volume element $d\tau = dx\,dy\,dz$ and $\nabla^2 = \partial^2/\partial x^2 + \partial^2/\partial y^2 + \partial^2/\partial z^2$, but it can be shown that

$$d\tau = r^2\,dr\,\sin\theta\,d\theta\,d\phi \tag{1-33}$$

and

$$\nabla^2 = \frac{1}{r^2}\frac{\partial}{\partial r}\left(r^2\frac{\partial}{\partial r}\right) + \frac{1}{r^2\sin\theta}\frac{\partial}{\partial\theta}\left(\sin\theta\frac{\partial}{\partial\theta}\right)$$

$$+ \frac{1}{r^2\sin^2\theta}\frac{\partial^2}{\partial\phi^2}. \tag{1-34}$$

If the atomic nucleus is placed in the center of the coordinate system with a positive charge of $Q|e|$, where $|e|$ is the numerical value of the charge of the electron, we obtain the Hamiltonian operator for an electron, mass m, in this central field:

$$\mathcal{H} = \frac{-\hbar^2}{2m} \nabla^2 - \frac{Qe^2}{r} . \tag{1-35}$$

The one electron eigenfunctions of this Hamiltonian operator are called *atomic orbitals*, and are designated $\chi(r,\theta,\phi)$.

With ∇^2 written out in polar coordinates, the Schrödinger equation for this system is

$$\left[\frac{1}{r^2} \frac{\partial}{\partial r} \left(r^2 \frac{\partial}{\partial r} \right) + \frac{1}{r^2 \sin \theta} \frac{\partial}{\partial \theta} \left(\sin \theta \frac{\partial}{\partial \theta} \right) \right.$$
$$\left. + \frac{1}{r^2 \sin^2 \theta} \frac{\partial}{\partial \phi^2} \right] \chi(r,\theta,\phi)$$
$$+ \frac{2m}{\hbar^2} \left[W + \frac{Qe^2}{r} \right] \chi(r,\theta,\phi) = 0. \tag{1-36}$$

As expected, the equation is a differential equation of second order in three variables. Let us try to guess a solution.

$$\chi(r,\theta,\phi) = R(r)Y(\theta,\phi). \tag{1-37}$$

Expression (1-37) is substituted into equation (1-36) and the r dependency is collected on the left side of the equation, the θ and ϕ dependency on the right:

$$\frac{1}{R(r)} \frac{\partial}{\partial r} \left(r^2 \frac{\partial R(r)}{\partial r} \right) + \frac{2m}{\hbar^2} \left(W + \frac{Qe^2}{r} \right) r^2$$
$$= - \frac{1}{Y(\theta,\phi) \sin \theta} \frac{\partial}{\partial \theta} \left(\sin \theta \frac{\partial Y(\theta,\phi)}{\partial \theta} \right)$$
$$- \frac{1}{Y(\theta,\phi) \sin^2 \theta} \frac{\partial^2 Y(\theta,\phi)}{\partial \phi^2} \tag{1-38}$$

Because r, θ, and ϕ are independent coordinates that can vary freely, each side can be equated to a constant, λ. Let Y stand for $Y(\theta,\phi)$ and R for $R(r)$. We obtain, then,

$$\frac{1}{\sin\theta}\frac{\partial}{\partial\theta}\left(\sin\theta\frac{\partial Y}{\partial\theta}\right) + \frac{1}{\sin^2\theta}\frac{\partial^2 Y}{\partial\phi^2} + \lambda Y = 0, \quad (1\text{-}39)$$

$$\frac{1}{r^2}\frac{d}{dr}\left(r^2\frac{dR}{dr}\right) + \left[\frac{2m}{\hbar^2}\left(W + \frac{Qe^2}{r}\right) - \frac{\lambda}{r^2}\right]R = 0.$$
$$(1\text{-}40)$$

We have split the Schrödinger equation into two independent second-order differential equations, one that depends only on the angles θ and ϕ and one that depends only on r. There are a number of solutions that satisfy the prescribed conditions (continuity and "good behavior" for $r \to \infty$), but a requirement for such solutions is that $\lambda = l(l+1)$, where l is zero or an integer: $l = 0,1,2,3,\ldots$.

For $l = 0$, $Y =$ constant is easily seen to be a solution of the angular equation. Normalizing this solution to one over the solid angle $\sin\theta\, d\theta\, d\phi$, we get $Y = (1/4\pi)^{1/2}$. Looking now at the radial equation for large values of r, (1-40) reduces to the asymptotic equation

$$\frac{d^2R}{dr^2} + \frac{2m}{\hbar^2}WR = 0.$$

Calling $(2m/\hbar^2)W = -\alpha^2$, we see that $R = \exp(\pm\alpha r)$ is a solution. We must discard the plus sign here, since for $r \to \infty$ the wavefunction would be ill-behaved; it cannot be normalized.

We now try by substitution to see if the asymptotic solution is also a solution to the complete radial differential equation with $l = 0$, and find this to be so provided that $\alpha = Qme^2/\hbar^2 = Q/a_0$, where $a_0 = \hbar^2/me^2$, the Bohr radius.

We could also have used the variational principle to find α. With $\chi = \exp(-\alpha r)$, we obtain

$$\langle W(\alpha) \rangle = \frac{\int \chi \mathcal{H} \chi \, d\tau}{\int |\chi|^2 \, d\tau} = \frac{\hbar^2}{2m} \alpha^2 - Qe^2\alpha. \qquad (1\text{-}41)$$

Minimizing $\langle W \rangle$ with respect to α gives again $\alpha = Q/a_0$.

The first normalized solution to the radial equation with $l = 0$ is then found from the condition

$$\int_0^\infty N^2 \exp(-2Qr/a_0) \, r^2 \, dr = 1$$

and is

$$R_{1,0}(r) = 2\left(\frac{Q}{a_0}\right)^{3/2} \exp(-Qr/a_0). \qquad (1\text{-}42)$$

All orbitals with $l = 0$ are called s orbitals, and the subscript $1,0$ associated with $R(r)$ refers to this function as the first radial solution to the complete set of s orbitals. The next one (similarly with $l = 0$) will be $R_{2,0}(r)$, and so on. The total wavefunction is the product of the radial and the angular function and we have

$$\chi_{1s} = \left(\frac{1}{\pi}\right)^{1/2} \left(\frac{Q}{a_0}\right)^{3/2} \exp(-Qr/a_0) \qquad (1\text{-}43)$$

The energy W_1 of an electron in a $1s$ orbital is found by substituting the value of α into the expression containing W:

$$-\alpha^2 = \frac{2W_1}{a_0 e^2} = \frac{Q^2}{a_0^2}. \qquad (1\text{-}44)$$

We obtain $W_1 = -Q^2 e^2/2a_0$.

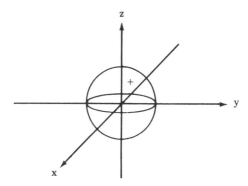

Figure 1-4 A contour surface of an *s* orbital; an *s* orbital has no angular dependency.

This is the most stable orbital of a hydrogen-like atom. By the most stable, we mean the orbital with the lowest energy. Since a 1*s* orbital has no angular dependency, the probability density $|\chi|^2$ is spherically symmetric. Furthermore, this is true for all *s* orbitals. We depict the contour surface[2] for an electron in an *s* orbital as a sphere (Figure 1-4).

The probability of finding a 1*s* electron within a spherical shell with radii r and $r + dr$ is found from (1-43) to be equal to

$$4\left(\frac{Q}{a_0}\right)^3 \exp(-2Qr/a_0)\, r^2\, dr. \qquad (1\text{-}45)$$

---

[2]The contour is a surface at which the wavefunction has the same numerical value. Whether the value is plus or minus is sometimes indicated. The contour is usually chosen to contain such a volume that on integration of $|\psi|^2$ a large fraction (for instance 90%) of the charge associated with the electron is found inside the surface.

By differentiation the above function is found to have a maximum at $r = a_0/Q$. At this distance (which for a hydrogen atom equals a_0) the electron density is largest and the chance of finding an electron is, therefore, the greatest.

To find $R_{j,0}$ a trial function of the form $f(r) \exp(-\beta r)$ is substituted into the radial differential equation (1-40). The polynomial $f(r)$ ensures that $R_{j,0}(r)$ can be made orthogonal to $R_{k,0}(r)$ ($k \neq j$). The radial solution $R_{2,0}(r)$ is given in Eq. (1-46), and the orthogonality of $R_{1,0}(r)$ and $R_{2,0}(r)$ can be inferred from Figure 1-5.

$$R_{2,0}(r) = \frac{1}{2(2)^{1/2}} \left(\frac{Q}{a_0} \right)^{3/2} \left(2 - \frac{Q}{a_0} r \right) \exp(-Qr/2a_0).$$

$$(1\text{-}46)$$

The total wavefunction is of course obtained by multipli-

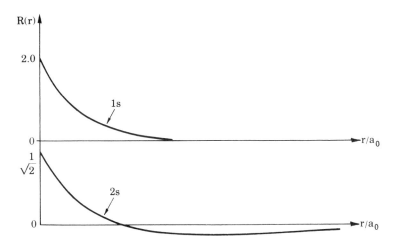

Figure 1-5 The behavior of the radial functions for the first two
s orbitals.

cation of $R_{2,0}(r)$ and the "angular part" $(4\pi)^{-1/2}$. This orbital is called a 2s orbital.

$$\chi_{2s} = \frac{(2\pi)^{-1/2}}{4} \left(\frac{Q}{a_0}\right)^{3/2} \left(2 - \frac{Q}{a_0} r\right) \exp(-Qr/2a_0).$$
$$(1\text{-}47)$$

The 2s orbital has the same angular dependency as 1s; owing to the node in the radial function, however, the probability distribution has two maxima, of which the larger lies farther out than the one that was found for 1s. The energy of 2s, W_2, is $-Q^2e^2/8a_0$.

The general expression for the energy is given by the Bohr formula

$$W_n = - \frac{Q^2e^2}{2n^2a_0}, \qquad n = 1,2,3,\ldots. \qquad (1\text{-}48)$$

We call n the principal quantum number, and n takes integer values $1,2,3,\ldots$. The exact solution of the Schrödinger equation for the hydrogen-like atom also requires $l \leq n-1$.

The energy difference between the 1s and 2s orbitals is found to be

$$\Delta W = W_{2s} - W_{1s} = \frac{Q^2e^2}{2a_0}\left(-\frac{1}{4} + 1\right) = \frac{3Q^2e^2}{8a_0}$$

which for $Q = 1$ gives $\Delta W = 3e^2/8a_0 \approx 10$ electron volts, abbreviated 10 eV.

We now turn to the possible solutions of the Schrödinger equation for $l \neq 0$. For $l = 1$ the angular equation gives three solutions that are orthogonal to each other. These are the so-called p orbitals. Unnormalized these are

$$p_z = \cos\theta = \frac{z}{r}, \qquad (1\text{-}49)$$

$$p_y = \sin\theta\,\sin\phi = \frac{y}{r}\,, \qquad (1\text{-}50)$$

$$p_x = \sin\theta\,\cos\phi = \frac{x}{r}\,. \qquad (1\text{-}51)$$

The subscripts x, y, and z indicate the angular dependencies. As expected, the three p orbitals are othogonal to each other, and it is obvious that they are not spherically symmetrical about the nucleus. A contour surface for each of the three p orbitals is given in Figure 1-6. Each set of p orbitals also has a radial function; that associated with the $2p$ set of functions is given in (1-52):

$$R_{2,1} = \frac{(6)^{-1/2}}{2}\left(\frac{Q}{a_0}\right)^{5/2} r\,\exp(-Qr/2a_0). \quad (1\text{-}52)$$

The energy of the $2p$ wavefunction is found to be the same as $2s$; that is, $W_2 = -(Q^2e^2/8a_0)$.

For $l = 2$ we obtain the following five linearly independent solutions to the angular equation; they are called the d orbitals:

$$d_{x^2-y^2}\colon\ \frac{(3)^{1/2}}{2}\,\sin^2\theta\,(\cos^2\phi - \sin^2\phi) = \frac{(3)^{1/2}}{2}\left(\frac{x^2 - y^2}{r^2}\right),$$
$$(1\text{-}53)$$

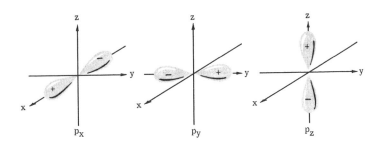

Figure 1-6 Contour surfaces of the p orbitals.

$$d_{z^2}: \quad \tfrac{1}{2}(3\cos^2\theta - 1) = \tfrac{1}{2}\left(\frac{3z^2 - r^2}{r^2}\right), \qquad (1\text{-}54)$$

$$d_{xy}: \quad (3)^{1/2}\sin^2\theta\cos\phi\sin\phi = (3)^{1/2}\,\frac{xy}{r^2}\,, \qquad (1\text{-}55)$$

$$d_{xz}: \quad (3)^{1/2}\sin\theta\cos\theta\cos\phi = (3)^{1/2}\,\frac{xz}{r^2}\,, \qquad (1\text{-}56)$$

$$d_{yz}: \quad (3)^{1/2}\sin\theta\cos\theta\sin\phi = (3)^{1/2}\,\frac{yz}{r^2}\,. \qquad (1\text{-}57)$$

The contour surfaces for the different d orbitals are given in Figure 1-7.

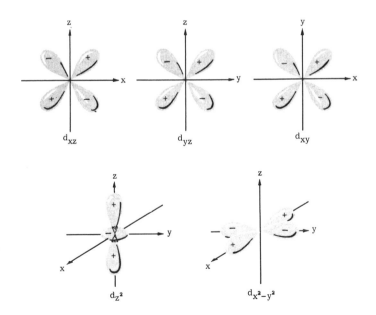

Figure 1-7 Contour surfaces of the d orbitals.

The radial function for a $3d$ orbital is

$$R_{3,2} = \frac{4(30)^{-1/2}}{81} \left(\frac{Q}{a_0} \right)^{7/2} r^2 \exp(-Qr/3a_0) \quad (1\text{-}58)$$

with energy $W_3 = -Q^2 e^2 / 18 a_0$.

1-4 THE ELECTRON SPIN

The electron may be viewed as a small ball; in addition to its orbital motion in (x,y,z) space, it "spins" around its axis. This motion introduces an additional degree of freedom that may be characterized by a quantum number. The electron spin was discovered from experiments that showed that an electron interacts with a magnetic field. Therefore, it has a magnetic moment. It was found that the spinning electron has only two quantum states, and that these states are degenerate in the absence of a magnetic field.

The wavefunctions for the two degenerate states are called $\alpha(s)$ and $\beta(s)$ where s is the spin coordinate. $\alpha(s)$ and $\beta(s)$ form an orthonormal set, and we have the normalizing conditions

$$\int |\alpha|^2 \, ds = \int |\beta|^2 \, ds = 1 \quad (1\text{-}59)$$

and the orthogonality condition

$$\int \alpha\beta \, ds = 0. \quad (1\text{-}60)$$

The total wavefunction for one electron is therefore given as a product of a space and spin-function, a so-called *spin orbital*.

$$\psi_i(x,y,z) \cdot (\alpha \text{ or } \beta).$$

A one electron spin-orbital is usually written as (ψ_i^α) or (ψ_i^β). It may also be called (ψ_i^+) or (ψ_i^-). Notice that no spin coordinate occurs in the electronic Hamilton function (1-35). As we shall see, the importance of the spin-orbitals is that for many-electron systems they help us to construct the proper wave-functions.

Chapter 1

Suggested Reading

V. Rojansky, Introductory Quantum Mechanics, Prentice-Hall, Englewood Cliffs, N.J., 1938.
Good for the basic principles.
M. Karplus and R. N. Porter, Atoms and Molecules, Benjamin, New York, 1970.
Instructive and sound treatment.

Problems

1. Calculate a reduced form of each of the following operators:

$$\hat{O} = \left[\frac{d}{dx} x - x \frac{d}{dx} \right],$$

$$\hat{O} = \left[\frac{d}{dx} x^2 - x^2 \frac{d}{dx} \right].$$

2. Is d/dx a Hermitian operator? Is $(1/i)(d/dx)$ a Hermitian operator? [*Hint.* Set up the integral $\int_{-\infty}^{\infty} \phi^*(x) \hat{O} \psi(x) \, dx$ and use partial integration, remembering the boundary conditions for the wavefunctions.]
3. The Hamilton operator for a harmonic oscillator is

$$\hat{H} = -\frac{\hbar^2}{2M} \frac{d^2}{dx^2} + \tfrac{1}{2}kx^2.$$

The x boundaries are $-\infty \le x \le \infty$; $M =$ mass of the oscillator; $k =$ force constant. The "classical" frequency is $\nu = (1/2\pi)\sqrt{k/M}$. Use the trial wavefunctions $\psi_0 = e^{-(\beta/2)x^2}$ and $\psi_1 = xe^{-(\beta/2)x^2}$ to calculate the two lowest energies. Prove that ψ_0 and ψ_1 are orthogonal to each other.

4. Use the Heisenberg uncertainty principle, $\delta\ddot{x}\,\delta p_x \geq \hbar/2$, to calculate the energy of a ball with mass M bouncing on a table under the influence of a gravitational field g.

5. Show that a $1s$ atomic orbital is orthogonal to $2s$. If the "effective nuclear charge" is different for $1s$ and $2s$ orbitals, are they still orthogonal? (Use the relationship $\int_0^\infty e^{-ax}x^n\,dx = n!/a^{n+1}$.)

MOLECULAR ORBITALS

2-1 DIATOMIC MOLECULES

Consider now the situation in which we have two positive nuclei with charges $+Q_A|e|$ and $+Q_B|e|$. Into this skeleton one electron is placed.

We have to solve the electronic Schrödinger equation, keeping the nuclei A and B fixed in space at a separation distance R. In the Schrödinger equation for the motion of an electron

$$\mathcal{H}\,\psi_i = W_i\psi_i \qquad (2\text{-}1)$$

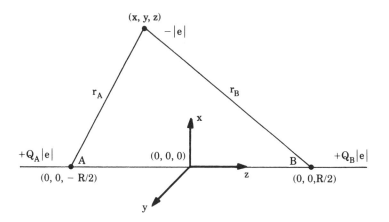

Figure 2-1 Molecular coordinate system for a diatomic molecule.

27

the Hamiltonian for a diatomic one-electron system is given by

$$\mathcal{H} = -\frac{\hbar^2}{2m} \left(\frac{\partial^2}{\partial x^2} + \frac{\partial^2}{\partial y^2} + \frac{\partial^2}{\partial z^2} \right)$$

$$- \frac{Q_A e^2}{\sqrt{x^2 + y^2 + (z + R/2)^2}} - \frac{Q_B e^2}{\sqrt{x^2 + y^2 + (z - R/2)^2}} \,.$$

$$(2\text{-}2)$$

The solutions, ψ_i, are called the *molecular orbitals* for a diatomic molecule.

If the electron is in the neighborhood of A, we expect that the atomic orbital χ_A will give a good description of its behavior. In the same way, the atomic orbital χ_B should give a good description of the behavior of the electron in the neighborhood of B. We then guess that a linear combination of χ_A and χ_B can be used as a trial wavefunction for the molecule AB. Thus, in the spirit of the variational principle, and assuming for convenience that χ_A and χ_B are real functions, we write the following trial function:

$$\psi = c_1 \chi_A + c_2 \chi_B. \qquad (2\text{-}3)$$

Such a molecular orbital is called an LCAO-MO ("linear combination of atomic orbitals-molecular orbital").

Before we make any calculation of the energy we shall normalize our wavefunction to 1. Then

$$N^2 \int (c_1 \chi_A + c_2 \chi_B)(c_1 \chi_A + c_2 \chi_B) \, d\tau = 1$$

or

$$(c_1)^2 \int |\chi_A|^2 \, d\tau + (c_2)^2 \int |\chi_B|^2 \, d\tau + 2c_1 c_2 \int \chi_A \chi_B \, d\tau$$

$$= \frac{1}{N^2} \,. \qquad (2\text{-}4)$$

Assuming that the atomic orbitals χ_A and χ_B are normalized, we have immediately

$$(c_1)^2 + (c_2)^2 + 2c_1c_2 \int \chi_A\chi_B \, d\tau = \frac{1}{N^2}. \quad (2\text{-}5)$$

The definite integral $\int \chi_A \chi_B \, d\tau$ is called the *overlap* integral, S_{AB} (S for short), between the two functions χ_A and χ_B. The reason for this name is clear. For example, if both χ_A and χ_B are s orbitals, we obtain the picture shown in Figure 2-2. We see that the two functions "overlap" each other in the region between the two nuclei. Provided $\chi_A = \chi_B$, we find $0 \leq |S| \leq 1$.

By substituting S for the overlap integral in Eq. (2-5), we obtain

$$(c_1)^2 + (c_2)^2 + 2c_1c_2S = \frac{1}{N^2}. \quad (2\text{-}6)$$

The normalized wavefunction is then

$$\psi = \frac{1}{\sqrt{(c_1)^2 + (c_2)^2 + 2c_1c_2S}} \, (c_1\chi_A + c_2\chi_B). \quad (2\text{-}7)$$

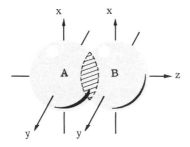

Figure 2-2 Overlap between two s functions. The cross-hatched area has electron density which originates from both χ_A and χ_B.

The energy of this trial wavefunction is given by the expression

$$\langle W \rangle = N^2 \int (c_1\chi_A + c_2\chi_B)\mathscr{H}(c_1\chi_A + c_2\chi_B)\, d\tau$$

$$= \frac{c_1^2 \int \chi_A\mathscr{H}\chi_A\, d\tau + c_2^2 \int \chi_B\mathscr{H}\chi_B\, d\tau + 2c_1c_2 \int \chi_A\mathscr{H}\chi_B\, d\tau}{c_1^2 + c_2^2 + 2c_1c_2 S}$$

$$(2\text{-}8)$$

where $\int \chi_B\mathscr{H}\chi_A\, d\tau = \int \chi_A\mathscr{H}\chi_B\, d\tau$ because $\mathscr{H}$ is a Hermitian operator. We now abbreviate the definite integrals as follows:

$$H_{AA} = \int \chi_A\mathscr{H}\chi_A\, d\tau, \qquad (2\text{-}9)$$

$$H_{BB} = \int \chi_B\mathscr{H}\chi_B\, d\tau, \qquad (2\text{-}10)$$

$$H_{AB} = H_{BA} = \int \chi_A\mathscr{H}\chi_B\, d\tau. \qquad (2\text{-}11)$$

Such integrals are called "matrix elements" of the orbitals χ_A and χ_B.

Since a definite integral is a number, we obtain $\langle W \rangle$ as a function of the parameters c_1 and c_2, and indirectly as a function of the chosen atomic ortibal exponents Z_A, Z_B, and R:

$$\langle W \rangle = \frac{(c_1)^2 H_{AA} + (c_2)^2 H_{BB} + 2c_1c_2 H_{AB}}{(c_1)^2 + (c_2)^2 + 2c_1c_2 S} \qquad (2\text{-}12)$$

We are interested in finding the values of c_1 and c_2, which for given atomic orbitals and constant R (that is, for constant H_{AA}, H_{BB}, and H_{AB}) will minimize the orbital energy $\langle W \rangle$. We use the variational principle to determine c_1 and c_2 by differentiating $\langle W \rangle$ with respect to these parameters.

From Eq. (2-12), we obtain

$$\langle W \rangle [(c_1)^2 + (c_2)^2 + 2c_1c_2S] = (c_1)^2H_{AA} + (c_2)^2H_{BB}$$
$$+ 2c_1c_2H_{AB} \qquad (2\text{-}13)$$

and by differentiation of (2-13) with respect to c_1,

$$\frac{\partial \langle W \rangle}{\partial c_1} [(c_1)^2 + (c_2)^2 + 2c_1c_2S] + \langle W \rangle (2c_1 + 2c_2S)$$
$$= 2c_1H_{AA} + 2c_2H_{AB}. \qquad (2\text{-}14)$$

Since the extremum value is found for $\partial \langle W \rangle /\partial c_1 = 0$, we immediately obtain an equation for determining the extremum values of W

$$c_1(H_{AA} - W) + c_2(H_{AB} - WS) = 0. \qquad (2\text{-}15)$$

The equation obtained from $\partial \langle W \rangle /\partial c_2$ can be written by inspection:

$$c_1(H_{AB} - WS) + c_2(H_{BB} - W) = 0. \qquad (2\text{-}16)$$

These two simultaneous, homogeneous equations have the solutions (not considering the trivial solution, $c_1 = c_2 = 0$) given by the following determinant:

$$\begin{vmatrix} H_{AA} - W & H_{AB} - WS \\ H_{BA} - WS & H_{BB} - W \end{vmatrix} = 0. \qquad (2\text{-}17)$$

This determinantal equation gives the extremum energy W of the "best" linear combination of χ_A and χ_B for the system. Expanding the equation gives

$$(H_{AA} - W)(H_{BB} - W) - (H_{AB} - WS)^2 = 0. \;(2\text{-}18)$$

A second-order equation has two roots, and we find the two values of W by solution of Eq. (2-18).

Let us discuss this equation. First we assume that $H_{AA} = H_{BB}$. In other words, we have a homonuclear diatomic molecule, where "both ends are equal". Examples are H_2, O_2, and Cl_2. Then

$$H_{AA} - W = \pm(H_{AB} - WS) \qquad (2\text{-}19)$$

or

$$W = \frac{H_{AA} \pm H_{AB}}{1 \pm S} . \qquad (2\text{-}20)$$

Since $|S|$ is less than 1 (if the A and B nuclei are together, $|S| = 1$; if they are infinitely separated, $S = 0$), we obtain one energy that is *smaller* than $H_{AA} = H_{BB}$ and one that is *larger*. These are called, respectively, the *bonding* and the *antibonding* solutions (see Figure 2-3). It is customary to designate the bonding solution as W^b and the antibonding solution as W^*. We see that the terms bonding and antibonding refer to the lowering or raising of the electronic energy when two atomic nuclei are brought together.

The sign of H_{AA} is negative, as is required for an electron bound to a positive nucleus. Calculations show that H_{AB} is also negative provided S is positive. Calling $H_{AB}/H_{AA} = \gamma$, we have $W = H_{AA}[(1\pm\gamma)/(1\pm S)]$. $W < H_{AA}$ (<0) for $(1\pm\gamma)/(1\pm S) > 1$ or $\pm\gamma > \pm S$. In other words if $\gamma > S$ the combination $\chi_A + \chi_B$ is bonding. This is always the case for diatomic molecules, but is not necessarily true for polyatomic molecules. Owing to the denominator $1 - S$ in the antibonding combination, we see that the antibonding orbital is destabilized more than the bonding orbital is stabilized (Figure 2-3).

The energy of the bonding (or antibonding) orbital is a function of R. Adding the nuclear repulsion term $Q_A Q_B e^2/R$ to the electronic energy, we can minimize W^b with respect to variations in R and obtain values for the energies as functions of R (Figure 2-4). Such curves are referred to as *potential functions*

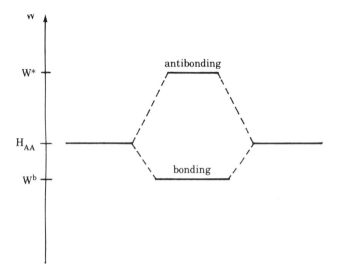

Figure 2-3   Energies of the bonding and antibonding orbitals in the AB molecule, pictured for a fixed value of R. The antibonding level is placed higher above H_{AA} than the bonding level is placed below.

for the molecule. Better numerical results can be obtained by also minimizing W with respect to the atomic orbital exponents, Z_{eff}.

For the H_2^+ system with $Z_{\text{eff}} = 1$, a minimum in the energy $(-1.76\ \text{eV})$ is found at $R = 1.3\ \text{Å}$. Therefore the latter value is the calculated equilibrium distance corresponding to a dissociation energy (D_e) of 1.76 eV. The experimental values are 1.06 Å and 2.79 eV. The important point emerging from this calculation (which can be greatly improved) is that the H_2^+ system is more stable than a separated H atom and a proton, H^+. In other words, a "chemical bond" has joined the two protons. Notice also that only the lowest potential function has a minimum. For this reason, the H_2^+ ion will dissociate in its first excited state.

We next turn to the determination of the coefficients c_1 and c_2 in our LCAO wavefunction (for $H_{AA} = H_{BB}$). This is done by

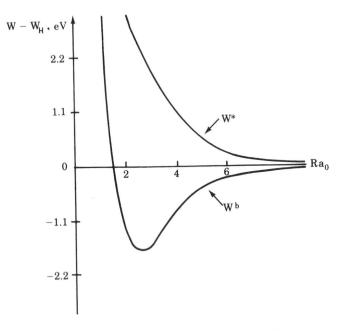

Figure 2-4   Lowest calculated energy levels for the H_2^+ molecule as a function of R. The energy minimum in the W^b potential curve occurs at $R = 1.3$ Å with $D_e = 1.76$ eV:

substituting the values found for W in the original set of equations (2-15) and (2-16):

$$c_1 = -c_2\left(\frac{H_{AB} - WS}{H_{AA} - W}\right), \qquad (2\text{-}21)$$

$$c_1 = -c_2\left(\frac{H_{AA} - W}{H_{AB} - WS}\right). \qquad (2\text{-}22)$$

Using (2-18) with $H_{AA} = H_{BB}$,

$$(H_{AB} - WS)^2 = (H_{AA} - W)^2 \qquad (2\text{-}23)$$

or

$$(H_{AB} - WS) = \pm(H_{AA} - W) \qquad (2\text{-}24)$$

we get by substitution in (2-21) and (2-22)

$$c_2 = -c_1 \quad \text{and} \quad c_2 = c_1. \qquad (2\text{-}25)$$

The two normalized wavefunctions are therefore

$$\frac{1}{\sqrt{2 + 2S}} \; (\chi_A + \chi_B) \;\; \text{(low energy)}, \qquad (2\text{-}26)$$

$$\frac{1}{\sqrt{2 - 2S}} \; (\chi_A - \chi_B) \;\; \text{(high energy)}. \qquad (2\text{-}27)$$

The coefficients c_1 and c_2 are in this case determined by the condition $H_{AA} = H_{BB}$. This is actually a *symmetry condition*, that is, a condition imposed because the molecule contains equivalent nuclei.

Let us now examine the general case given by Eq. (2-18):

$$(H_{AA} - W)(H_{BB} - W) - (H_{AB} - WS)^2 = 0. \qquad (2\text{-}28)$$

With H_{AB} numerically smaller than $|H_{AA} - H_{BB}|$, we obtain an approximate solution of (2-28), by setting $W = H_{AA}$ or H_{BB} whenever the term does not become zero. Then

$$(H_{AA} - W)(H_{BB} - H_{AA}) \approx (H_{AB} - H_{AA}S)^2 , \qquad (2\text{-}29a)$$

$$(H_{AA} - H_{BB})(H_{BB} - W) \approx (H_{AB} - H_{BB}S)^2 , \qquad (2\text{-}29b)$$

or

$$W \approx \begin{cases} H_{AA} - \dfrac{(H_{AB} - SH_{AA})^2}{H_{BB} - H_{AA}} , \\[4mm] H_{BB} + \dfrac{(H_{AB} - SH_{BB})^2}{H_{BB} - H_{AA}} . \end{cases} \qquad (2\text{-}30)$$

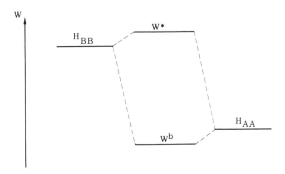

Figure 2-5   Bonding and antibonding energy levels for $H_{BB} \gg H_{AA}$.

H_{AA} and H_{BB} are the energies of the atomic orbitals χ_A and χ_B, respectively, in the molecular skeleton. One solution will therefore give an orbital energy lower than H_{AA}, whereas the other one will be higher in energy than H_{BB} (Figure 2-5). Since $H_{AB} \approx SH_{BB}$, Eq. (2-30) tells us that the bonding energy is proportional to $S^2(H_{AA} - H_{BB})$. To have strong bonding, then, the overlap between the valence atomic orbitals in question must be large.

With S given by the integral $\int \chi_A \chi_B \, d\tau$, we shall now show that for S to be different from zero the atomic orbitals χ_A and χ_B in Eq. (2-3) must have the *same symmetry* around the line between A and B. Consider, for example, the overlap between an s and a p_y orbital. Figure 2-6 shows that each small volume element "in the top" of the p_y orbital is positive, whereas the corresponding volume element "below" is negative. Therefore, a summation (the integration) over all volume elements will give zero in the case under consideration,

$$\int (s)_A (p_y)_B \, d\tau = 0. \qquad (2-31)$$

The overlaps of an s orbital on A with various orbitals on B are shown in Figure 2-7. We have $\int (s)_A (p_z)_B \, d\tau \neq 0$,

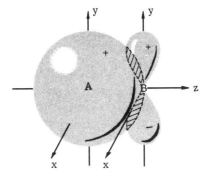

Figure 2-6   Overlap between an s and a p_y orbital.

$\int (s)_A (d_{z^2})_B \, d\tau \neq 0$, but $\int (s)_A (d_{x^2-y^2})_B \, d\tau = 0$. Thus, the general rule emerges that the overlap integral is zero if the two orbitals involved have different symmetries around the connecting axis.

We shall next show that the symmetry conditions that cause an overlap integral to disappear also require that $H_{AB} = \int \chi_A \mathscr{H} \chi_B \, d\tau$ be zero. The Hamiltonian for a diatomic molecule with the nuclear repulsion included is

$$\mathscr{H} = - \frac{\hbar^2}{2m} \nabla^2 - \frac{Q_A e^2}{r_A} - \frac{Q_B e^2}{r_B} + \frac{Q_A Q_B e^2}{R} . \quad (2\text{-}32)$$

We have, recalling that χ_A and χ_B are solutions to the atomic problem,

$$\mathscr{H} \chi_B = \left(- \frac{\hbar^2}{2m} \nabla^2 - \frac{Q_B e^2}{r_B} \right) \chi_B - \frac{Q_A e^2}{r_A} \chi_B$$

$$+ \frac{Q_A Q_B e^2}{R} \chi_B = W_B \chi_B - \frac{Q_A e^2}{r_A} \chi_B + \frac{Q_A Q_B e^2}{R} \chi_B.$$

$$(2\text{-}33)$$

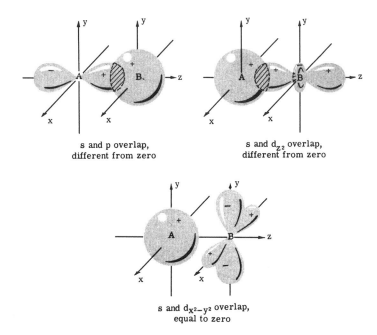

s and p overlap,
different from zero

s and d_{z^2} overlap,
different from zero

s and $d_{x^2-y^2}$ overlap,
equal to zero

Figure 2-7   Some different overlaps between an s orbital and p and d orbitals.

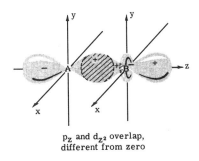

p_z and d_{z^2} overlap,
different from zero

Figure 2-8 p_z and d_{z^2} overlap, different from zero.

Thus,

$$\int \chi_A \mathscr{H} \chi_B \, d\tau = \left(W_B + \frac{Q_A Q_B e^2}{R} \right) S - Q_A e^2 \int \frac{\chi_A \chi_B}{r_A} \, d\tau.$$

$$(2\text{-}34)$$

If χ_A and χ_B have different symmetrics about $A-B$, S is zero. The integral $\int (\chi_A \chi_B / r_A) \, d\tau$ is also zero, since every volume element, divided by the distance from A, has a corresponding volume element of opposite sign, also divided by the same distance to A. H_{AB} follows S; both are zero if the two wavefunctions, centered respectively on A and B, do not have the same symmetry around the line joining the nuclei.

We distinguish among various types of molecular orbitals by means of "symmetry" of the overlap. If the overlap is symmetric for rotation around the "bond axis" $A-B$ (as, for example, $s_A - s_B$, $s_A - p_{zB}$, $p_{zA} - p_{zB}$, $p_{zA} - d_{z^2B}$, Figures 2-2, 2-7, and 2-8, etc.), the resulting molecular orbitals are called sigma (σ) orbitals. If the overlap leads to a nodal plane along the internuclear axis ($p_{xA} - p_{xB}$, $p_{yA} - p_{yB}$), Figure 2-9, the resulting molecular orbitals are called pi (π) orbitals.

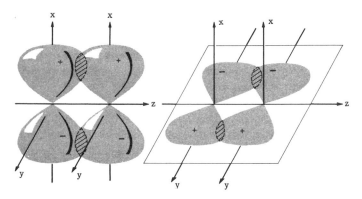

Figure 2-9 π-type overlap.

Note that the three p orbitals on each of two atoms give one σ bond (which uses the p_z orbital on each atomic nucleus) and two π bonds (which use $p_{xA} - p_{xB}$ and $p_{yA} - p_{yB}$). The resulting π_x and π_y molecular orbitals have the same energy, since p_x and p_y are equivalent in the molecule. Furthermore, since for ordinary bond distances the S_π overlap is *smaller* than the S_σ overlap, we expect π bonding to be weaker than σ bonding. It also follows that the π antibonding orbitals have lower energy than the σ antibonding orbitals (Figure 2-10).

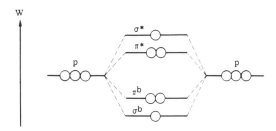

Figure 2-10   Bonding and antibonding molecular orbitals composed of the p orbitals in a homonuclear diatomic molecule.

Summarizing, we notice that the building up of electronic charge between two nuclei is equivalent to the formation of a chemical bond. Conversely, "antibonding" is associated with the presence of a nodal plane between the nuclei.

2-2   SYMMETRY OPERATORS

Consider the one-electron Hamiltonian for a diatomic molecule, Eq. (2-2)

$$\mathscr{H} = -\frac{\hbar^2}{2m} \left(\frac{\partial^2}{\partial x^2} + \frac{\partial^2}{\partial y^2} + \frac{\partial^2}{\partial z^2} \right)$$
$$- \frac{Q_A e^2}{\sqrt{x^2 + y^2 + (z + R/2)^2}} - \frac{Q_B e^2}{\sqrt{x^2 + y^2 + (z - R/2)^2}} \cdot$$

$$(2\text{-}35)$$

Suppose now that we change the electronic coordinates from a (x,y,z) system to a $(-x,-y,-z)$ system. We call such an operation an *inversion* of the coordinate system and write formally $\hat{i}(x,y,z) = (-x,-y,-z)$. Operating with $\hat{i}$ on $\mathscr{H}$, we obtain

$$\hat{i}\mathscr{H} = -\frac{\hbar^2}{2m}\left(\frac{\partial^2}{\partial x^2} + \frac{\partial^2}{\partial y^2} + \frac{\partial^2}{\partial z^2}\right)$$
$$-\frac{Q_A e^2}{\sqrt{x^2 + y^2 + (-z + R/2)^2}} - \frac{Q_B e^2}{\sqrt{x^2 + y^2 + (-z - R/2)^2}}.$$

$$(2\text{-}36)$$

Notice that only when $Q_A = Q_B$ is the transformed Hamiltonian equal to the original. Therefore, if Q_A equals Q_B (that is, for a homonuclear diatomic molecule), the inversion operator $\hat{i}$ does not change the molecular Hamiltonian. In such a case, we say that the molecule possesses a center of inversion (or center of symmetry).

In general, all operators that leave the molecular Hamiltonian function unchanged are called *symmetry operators*. For a nondegenerate orbital ψ and a symmetry operator $\hat{O}$, we have

$$\mathscr{H}\psi = W\psi \qquad (2\text{-}37)$$

and

$$\hat{O}(\mathscr{H}\psi) = \hat{O}(W\psi) \qquad (2\text{-}38)$$

or

$$\mathscr{H}(\hat{O}\psi) = W(\hat{O}\psi). \qquad (2\text{-}39)$$

Evidently, both $(\hat{O}\psi)$ and ψ are solutions to the same Schrödinger equation. They can, therefore, only differ by a constant, and we have

$$(\hat{O}\psi) = c\psi. \qquad (2\text{-}40)$$

Let, for instance, $\hat{O}$ be the inversion operator $\hat{i}$. By performing the inversion operation and then repeating it, we get

$$\hat{i}\hat{i}\psi = (\hat{i})^2\psi = \psi$$
$$\parallel$$
$$\hat{i}c\psi = c\hat{i}\psi = c^2\psi$$

It follows that $c^2 = 1$, or $c = \pm 1$.

Therefore, provided the molecule has a center of symmetry, the wavefunction must go into plus itself or into minus itself by applying $\hat{i}$. In that case we can characterize the wavefunction by g if $\hat{i}\psi_g = 1\psi_g$ and u if $\hat{i}\psi_u = -1\psi_u$, where g and u stand for even and odd (gerade and ungerade in German). $+1$ and -1 are seen to be the eigenvalues of the symmetry operator $\hat{i}$.

For any molecule there are a certain number of symmetry operations that leave the molecular Hamiltonian unaltered. These are operations that rotate, reflect, or invert the molecule into a configuration that is indistinguishable from the original one. One symmetry operation is always present, of course, and that is $\hat{E}$, the so-called identity, the application of which does nothing at all to the molecule.

The eigenfunctions of a molecular Hamiltonian can be characterized by their transformation properties under the symmetry operations of the molecule. The eigenvalues of the symmetry operators represent what are normally called *good quantum numbers*; each set of eigenvalues is given a characteristic designation. The importance of this lies in the fact that even though one is dealing with trial wavefunctions, such wavefunctions can be characterized exactly.

A $\hat{C}_n$ symmetry operation is defined as a rotation around an axis by an angle $2\pi/n$; this operation brings the nuclei of the molecule into an orientation that is equivalent to the original one. By applying it n-times we must have

$$(\hat{C}_n)^n = \hat{E}. \tag{2-41}$$

In the case of a nondegenerate molecular orbital, ψ, we must, as before, have

$$(\hat{C}_n\psi) = c\psi. \qquad (2\text{-}42)$$

It follows that

$$(c)^n = 1. \qquad (2\text{-}43)$$

For an even value of n the eigenvalue of a nondegenerate orbital under $\hat{C}_n$ is $+1$ or -1. If n is odd, only $+1$ is an eigenvalue.

The situation is more complicated in the case of electronic degeneracy. Consider the case of a two-fold degenerate state with wavefunctions ψ_1 and ψ_2. Because any linear combination of ψ_1 and ψ_2 have the same energy, W, in general we have

$$\hat{C}_n\psi_1 = a_{11}\psi_1 + a_{12}\psi_2 \qquad a_{11}^2 + a_{12}^2 = 1 \quad (2\text{-}44)$$

and

$$\hat{C}_n\psi_2 = a_{21}\psi_1 + a_{22}\psi_2 \qquad a_{21}^2 + a_{22}^2 = 1. \quad (2\text{-}45)$$

Owing to orthogonality $a_{11}a_{21} + a_{12}a_{22} = 0$.
Written in matrix language

$$\hat{C}_n\begin{pmatrix}\psi_1\\\psi_2\end{pmatrix} = \begin{pmatrix}a_{11} & a_{12}\\a_{21} & a_{22}\end{pmatrix}\begin{pmatrix}\psi_1\\\psi_2\end{pmatrix}. \qquad (2\text{-}46)$$

The sum of the diagonal elements of a matrix is known as *the character* of the matrix. We have, consequently, that the character of the transformation matrix in (2-46) equals $a_{11} + a_{22}$. If a symmetry operation of the molecule is, e.g., a rotation of $+2\pi/3$, a rotation by $-2\pi/3$ is also a symmetry operation. However, a mathematical theorem tells us that for such *equivalent operations*, the character of the transformation matrix will remain the same. For a degenerate set of orbitals, the character of the transformation matrix takes the place of the constant c in Eq. (2-42); indeed, c could be viewed as the character of a one-by-one matrix. The set of characters under the various symmetry operations are used to characterize the wavefunctions.

To illustrate the above points, we shall consider a hypothetical molecule with three equivalent nuclei placed at the corners of an equilateral triangle. As molecular orbitals we take linear combinations of three equivalent atomic orbitals χ_i centered on nuclei one, two, and three.

$$\psi = c_1\chi_1 + c_2\chi_2 + c_3\chi_3. \qquad (2\text{-}47)$$

Obviously, the molecule has a three-fold rotation axis. Since for a nondegenerate orbital

$$\hat{E}\psi = \psi = (\hat{C}_3)^3\psi = (\hat{C}_3)^2\lambda\psi = \lambda^3\psi \qquad (2\text{-}48)$$

we have $1 = \lambda^3$ or $\lambda = 1$. In other words, a nondegenerate orbital must remain unchanged when rotated $2\pi/3$. Therefore

$$\hat{C}_3(c_1\chi_1 + c_2\chi_2 + c_3\chi_3) \equiv c_1\chi_2 + c_2\chi_3 + c_3\chi_1$$
$$= 1 \cdot (c_1\chi_1 + c_2\chi_2 + c_3\chi_3). \qquad (2\text{-}49)$$

The only possible solution to (2-49) is $c_1 = c_2 = c_3$.

The nondegenerate orbital ψ_1 is then

$$\psi_1 = c_1(\chi_1 + \chi_2 + \chi_3). \qquad (2\text{-}50)$$

If we set the overlap integrals equal to zero,

$$\int \chi_1\chi_2 \, d\tau = \int \chi_2\chi_3 \, d\tau = \int \chi_1\chi_3 \, d\tau = 0 \qquad (2\text{-}51)$$

we obtain a normalized ψ_1 orbital:

$$\psi_1 = \sqrt{\tfrac{1}{3}}\,(\chi_1 + \chi_2 + \chi_3). \qquad (2\text{-}52)$$

Out of a basis set of three atomic orbitals we can construct three orthogonal molecular orbitals. The combination $(\chi_2 - \chi_3)$ is obviously orthogonal to ψ_1; normalizing to one, we have $\psi_2 = \sqrt{\tfrac{1}{2}}(\chi_2 - \chi_3)$. We have $\sqrt{\tfrac{2}{3}}$ of orbital one and $\sqrt{\tfrac{1}{6}}$ of each of orbitals two and three "left over". Evidently, $\psi_3 = \sqrt{\tfrac{1}{6}}(2\chi_1 - \chi_2 - \chi_3)$ is acceptable for the third combination. We know that ψ_2 and ψ_3 cannot be nondegenerate orbitals. We test this as follows:

$$\hat{C}_3 \sqrt{\tfrac{1}{6}}(2\chi_1 - \chi_2 - \chi_3) \equiv \sqrt{\tfrac{1}{6}}(2\chi_2 - \chi_3 - \chi_1). \qquad (2\text{-}53)$$

This is not one of the two basis combinations but it can be decomposed.

$$\sqrt{\tfrac{1}{6}}(2\chi_2 - \chi_3 - \chi_1) = -\tfrac{1}{2}\sqrt{\tfrac{1}{6}}(2\chi_1 - \chi_2 - \chi_3)$$
$$+ \frac{\sqrt{3}}{2}\sqrt{\tfrac{1}{2}}(\chi_2 - \chi_3). \qquad (2\text{-}54)$$

Further

$$\hat{C}_3\sqrt{\tfrac{1}{2}}(\chi_2 - \chi_3) \equiv \sqrt{\tfrac{1}{2}}(\chi_3 - \chi_1), \qquad (2\text{-}55)$$

$$\sqrt{\tfrac{1}{2}}(\chi_3 - \chi_1) = -\tfrac{1}{2}\sqrt{\tfrac{1}{2}}(\chi_2 - \chi_3) - \frac{\sqrt{3}}{2}\sqrt{\tfrac{1}{6}}(2\chi_1 - \chi_2 - \chi_3). \qquad (2\text{-}56)$$

Therefore, we can write for the doubly degenerate orbitals ψ_2 and ψ_3

$$\hat{C}_3 \begin{pmatrix} \psi_2 \\ \psi_3 \end{pmatrix} = \begin{pmatrix} -\tfrac{1}{2} & -\sqrt{\tfrac{3}{2}} \\ \sqrt{\tfrac{3}{2}} & -\tfrac{1}{2} \end{pmatrix} \begin{pmatrix} \psi_2 \\ \psi_3 \end{pmatrix}. \qquad (2\text{-}57)$$

The trace of the transformation matrix equals $-\tfrac{1}{2} + (-\tfrac{1}{2})$, or -1.

The orbital ψ_1 is designated a, whereas the doubly degenerate set of orbitals (ψ_2, ψ_3) is designated e. The degeneracies are indeed borne out by calculations. With the trial orbital (2-47), taking $H_{11} = H_{22} = H_{33} = \int \chi_1 \mathscr{H} \chi_1 \, d\tau$ and $H_{12} = H_{13} = H_{23} = \int \chi_1 \mathscr{H} \chi_2 \, d\tau$, we obtain the secular equation

$$\begin{vmatrix} H_{11} - W & H_{12} & H_{12} \\ H_{12} & H_{11} - W & H_{12} \\ H_{12} & H_{12} & H_{11} - W \end{vmatrix} = 0. \quad (2\text{-}58)$$

The solutions are found by expansion

$$W_1 = H_{11} + 2H_{12}, \quad (2\text{-}59)$$

$$\left. \begin{matrix} W_2 \\ W_3 \end{matrix} \right\} = H_{11} - H_{12}. \quad (2\text{-}60)$$

Using the symmetry-determined orbitals, however, we immediately obtain the solutions:

$$W(a) = \tfrac{1}{3} \int (\chi_1 + \chi_2 + \chi_3) \mathscr{H} (\chi_1 + \chi_2 + \chi_3)\, d\tau = H_{11} + 2H_{12},$$
$$(2\text{-}61)$$

$$W(e_1) = \tfrac{1}{2} \int (\chi_2 - \chi_3) \mathscr{H} (\chi_2 - \chi_3)\, d\tau = H_{11} - H_{12}, \quad (2\text{-}62)$$

$$W(e_2) = \tfrac{1}{6} \int (2\chi_1 - \chi_2 - \chi_3) \mathscr{H} (2\chi_1 - \chi_2 - \chi_3)\, d\tau = H_{11} - H_{12}.$$
$$(2\text{-}63)$$

Based on the symmetry-determined orbitals, the calculation of the LCAO-MO energies pertinent to an equilateral triangular molecular skeleton does not require a formal minimization procedure. Whenever the symmetry of the molecule fixes the coefficients in the molecular orbitals, an analogous procedure will work.

We shall now state a powerful theorem concerning molecular integrals involving symmetry-characterized functions: *Any molecular integral is zero unless the function to be integrated is totally symmetric under all symmetry operations relevant to the molecule.* A "totally symmetric" function is defined as a function that goes into plus itself under all relevant symmetry operations.

We can easily make this plausible by induction. Let us assume that f is not totally symmetric. Then by definition there will be some symmetry operation $\hat{O}$ that makes f change sign. If we now apply $\hat{O}$ to the integral $\int f\, d\tau$, we obtain:

$$\hat{O} \int f\, d\tau = \int (-f)\, d\tau = -\int f\, d\tau. \quad (2\text{-}64)$$

However, the application of a symmetry operation is equivalent to a change in the choice of coordinate system for the molecule, and since the value of a definite integral cannot depend on an arbitrary choice of coordinate system, we have

$$A = \int f \, d\tau = \hat{O} \int f \, d\tau = -\int f \, d\tau = -A. \quad (2\text{-}65)$$

The integral is equal to minus itself and must vanish.

This theorem immediately tells us that two molecular orbitals that transform differently are orthogonal to each other. Consider two orbitals that transform differently, $\psi_1(x,y,z)$ and $\psi_2(x,y,z)$. The overlap integral is

$$\int \psi_1^*((x,y,z)\psi_2(x,y,z) \, d\tau.$$

We observe that at least one symmetry operation will change the sign of one—but not the other ψ, since otherwise ψ_1 and ψ_2 would not transform differently. Consequently, the product $\psi_1^*\psi_2$ will not be totally symmetric and the integral must be zero by our theorem.

A matrix element $\int \psi_1^* \mathscr{H} \psi_2 \, d\tau$ is also zero if ψ_1 and ψ_2 transform in different ways. This follows because $\mathscr{H}$ is always a totally symmetric function, and the product $\psi_1^* \mathscr{H} \psi_2$ is therefore a nontotally symmetric function.

We can use this feature to improve our variational functions. Consider a molecule in which two molecular orbitals ψ_0 and ψ_1 have the energies w_0 and w_1 ($w_0 < w_1$), respectively. We write an augmented variational wavefunction

$$\psi = c_0\psi_0 + c_1\psi_1. \quad (2\text{-}66)$$

The energy of this variational orbital is found by solving

$$\begin{vmatrix} w_0 - w & H_{01} - S_{01}w \\ H_{01} - S_1 w & w_1 - w \end{vmatrix} = 0. \quad (2\text{-}67)$$

Here,

$$w_0 = \int \psi_0 \mathscr{H} \psi_0 \, d\tau, \qquad H_{01} = \int \psi_0 \mathscr{H} \psi_1 \, d\tau,$$

$$w_1 = \int \psi_1 \mathscr{H} \psi_1 \, d\tau, \quad \text{and} \quad S_{01} = \int \psi_0 \psi_1 \, d\tau.$$

Notice that only provided H_{01} and S_{01} are different from zero will the energy of the ground state be lowered. Then, according to the variational principle, we would have an improved wavefunction. But for this to happen ψ_0 and ψ_1 *must transform in the same way under all symmetry operations,* because only in that case will the product $(\psi_0 \cdot \psi_1)$ transform as a totally symmetric function.

By expansion of (2-67) we find the improved molecular orbital energies

$$\tilde{w}_0^b \approx w_0 - \frac{(H_{01} - S_{01}w_0)^2}{w_1 - w_0} \qquad (2\text{-}68)$$

and

$$\tilde{w}^* \approx w_1 + \frac{(H_{01} - S_{01}w_1)^2}{w_1 - w_0}. \qquad (2\text{-}69)$$

Therefore, levels with the same symmetry act as if they repel each other. Picturing the energy levels as functions of some parameter (for instance, an interatomic distance), levels having the same symmetry will never be able to cross. On the other hand, when the levels have different symmetries, nothing prevents them from crossing each other.

Chapter 2

Suggested Reading

M. Karplus and R. N. Porter, Atoms and Molecules, Benjamin, New York, 1970.

F. A. Cotton, Chemical Applications of Group Theory, 2nd ed., Wiley-Interscience, New York, 1971.

A very good introduction to the subject.

Problems

1. The elliptical coordinates λ and μ are defined as follows (see figure):

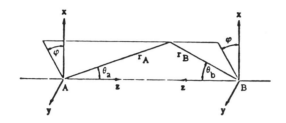

$$\lambda = \frac{r_A + r_B}{R} \, ,$$

$$\mu = \frac{r_A - r_B}{R} \, ,$$

$$d\tau = \frac{R^3}{8} (\lambda^2 - \mu^2)d\lambda \, d\mu \, d\varphi,$$

$$1 \leq \lambda \leq \infty, \qquad -1 \leq \mu \leq 1, \qquad 0 \leq \varphi \leq 2\pi.$$

R is the distance AB. Further,

$$r_A \cos \theta_a = \frac{R}{2} (1 + \lambda\mu),$$

$$r_B \cos \theta_b = \frac{R}{2} (1 - \lambda\mu),$$

$$r_A \sin \theta_a = r_B \sin \theta_b = \frac{R}{2} [(\lambda^2 - 1)(1 - \mu^2)]^{1/2}.$$

Calculate the overlap integral, $S(1s,1s)$, between two $1s$ orbitals, centered at a distance R from each other. What is the value of $S(1s,1s)$ for $R \to 0$ and ∞? For the $1s$ wavefunction, take

$$\chi_{1s} = \sqrt{\frac{1}{\pi}} \left(\frac{Z}{a_0}\right)^{3/2} \exp\left(\frac{-Zr_i}{a_0}\right) \qquad i = A \text{ or } B.$$

2. With the normalized $2p_z$ and $2p_x$ orbitals given by $2p_z = R(r) \cos \theta$ and $2p_x = R(r) \sin \theta \cos \varphi$, with

$$R(r) = \frac{1}{4\sqrt{2\pi}} \left(\frac{Z}{a_0}\right)^{5/2} r \exp\left(-\frac{Z}{2a_0}r\right)$$

calculate the following using elliptical coordinates (see Problem 1):

$$S(p_\sigma) = S(2p_zA, 2p_zB),$$

$$S(p_\pi) = S(2p_xA, 2p_xB) = S(2p_yA, 2p_yB).$$

The answers should be expressed using the so-called A integrals, defined as

$$A_n(\alpha) = \int_1^\infty x^n e^{-\alpha x}\, dx = \frac{n!\, e^{-\alpha}}{\alpha^{n+1}} \sum_{k=0}^n \frac{\alpha^k}{k!}.$$

Calculate numerical values for S_π and S_σ, with $Z = 4$ and $R = 2a_0$.

3. In H_2^+ we take

$$\chi_A = \chi_B = 1s = \sqrt{\frac{1}{\pi}\left(\frac{Z}{a_0}\right)^{3/2}} \exp\left(-\frac{Z}{a_0}r_i\right) \qquad i = A, B$$

and

$$\mathcal{H} = -\frac{\hbar^2}{2m}\nabla^2 - \frac{Ze^2}{r_A} - \frac{Ze^2}{r_B} + \frac{Z^2e^2}{R}.$$

Show that $H_{AA} = H_{BB}$, as well as $H_{AB} = H_{BA}$, using elliptical coordinates as in Problem 1. With the overlap integral from Problem 1, obtain the expression for the lowest orbital in H_2^+ as a function of R and Z. Putting $Z = 1$, sketch the curve $W = f(R)$.

4. Construct a molecular orbital energy level diagram for HF. Compare and contrast the bonding and the energy levels in HF and LiH.

MANY-ELECTRON SYSTEMS

3-1 THE PAULI PRINCIPLE

In this chapter we shall examine atoms and molecules containing more than one electron. In order to describe such systems we use atomic or molecular orbitals as building blocks. Consider the case of a two-electron system. The electronic Hamiltonian is

$$\mathscr{H} = \mathscr{H}_1 + \mathscr{H}_2 + e^2/r_{12} \qquad (3\text{-}1)$$

which is made up of the one-electron Hamiltonians

$$\mathscr{H}_1 = \frac{-\hbar^2}{2m} \left(\frac{\partial^2}{\partial x_1^2} + \frac{\partial^2}{\partial y_1^2} + \frac{\partial^2}{\partial z_1^2} \right) + V(x_1, y_1, z_1), \qquad (3\text{-}2)$$

$$\mathscr{H}_2 = \frac{-\hbar^2}{2m} \left(\frac{\partial^2}{\partial x_2^2} + \frac{\partial^2}{\partial y_2^2} + \frac{\partial^2}{\partial z_2^2} \right) + V(x_2, y_2, z_2) \qquad (3\text{-}3)$$

and the electronic repulsion term

$$\frac{e^2}{r_{12}} = \frac{e^2}{\sqrt{(x_1 - x_2)^2 + (y_1 - y_2)^2 + (z_1 - z_2)^2}}. \qquad (3\text{-}4)$$

The Schrödinger equation is

$$\mathscr{H}\Psi_k(1,2) = W_k\Psi_k(1,2). \qquad (3\text{-}5)$$

It is important to note that the Hamiltonian operator for the system, Eq. (3-1), remains unchanged upon performing the *permutation operation*, $P_{1,2}$. The permutation operator, $\hat{P}_{1,2}$, exchanges the electrons; electron 1 becomes electron 2, and electron 2 becomes electron 1, as follows:

$$\hat{P}_{1,2}\mathscr{H} = \hat{P}_{12}(\mathscr{H}_1 + \mathscr{H}_2 + e^2/r_{12}) = \mathscr{H}_2 + \mathscr{H}_1 + e^2/r_{21} = \mathscr{H}. \tag{3-6}$$

Operating with $\hat{P}_{1,2}$ on Eq. (3-5), therefore, gives

$$P_{1,2}(\mathscr{H}\Psi(1,2)) = \mathscr{H}(\hat{P}_{1,2}\Psi(1,2))$$

$$\| \tag{3-7}$$

$$\hat{P}_{1,2}(W\Psi(1,2)) = W(P_{1,2}\Psi(1,2)).$$

In other words, the wavefunction $(\hat{P}_{1,2}\Psi(1,2))$ is a solution to the Schrödinger equation as is $\Psi(1,2)$. From the Schrödinger equation, we see that the wavefunctions are only determined less a multiplier, hence

$$(\hat{P}_{1,2}\Psi(1,2)) = c\Psi(1,2). \tag{3-8}$$

Applying $\hat{P}_{1,2}$ two times in succession,

$$\hat{P}_{1,2}^2\Psi(1,2) = \Psi(1,2) = c^2\Psi(1,2) \tag{3-9}$$

or

$$c^2 = 1. \tag{3-10}$$

Therefore

$$c = \pm 1. \tag{3-11}$$

Under the *permutation operator* $\hat{P}_{1,2}$ the wavefunction either changes sign or remains the same. It turns out that calculations employing many-electron wavefunctions that do not change sign upon permutation of two electrons yield results that disagree with experiment. On the other hand, if the wavefunction changes sign upon permutation of the electrons, then the results are in accord with nature. Electronic wavefunctions, therefore, must change sign upon permutation of electronic coordinates; that is, *a many-electron wavefunction must by antisymmetric.* This statement is one formulation of the *Pauli principle.*

Taking account of the electronic spins, we can construct an antisymmetric *product wavefunction* for a two-electron system

$$\Psi(1,2) = \overset{\alpha}{\psi_j}(1) \overset{\beta}{\psi_j}(2) - \overset{\alpha}{\psi_j}(2) \overset{\beta}{\psi_j}(1). \qquad (3\text{-}12)$$

In this two-electron wavefunction each electron is in the same space orbital function found by solving the one-electron Schrödinger equation

$$\mathscr{H}_1\psi_j(1) = w_j\psi_j(1). \qquad (3\text{-}13)$$

However, the two electrons occupy different spin functions, viz.,

$$\overset{\alpha}{\psi}_j \quad \text{or} \quad \overset{\beta}{\psi}_j.$$

Integrating over both space and spin coordinates of electrons one and two, the normalization constant is found to be $\sqrt{\tfrac{1}{2}}$.

$$\Psi(1,2) = \sqrt{\tfrac{1}{2}} \, [\overset{\alpha}{\psi_j}(1) \overset{\beta}{\psi_j}(2) - \overset{\alpha}{\psi_j}(2) \overset{\beta}{\psi_j}(1)]. \qquad (3\text{-}14)$$

We see that if a product wavefunction is to be antisymmetric two electrons in any given space orbital cannot be in the same spin orbital, or, in what amounts to the same thing, only two electrons can occupy the same space orbital. This is another formulation of the Pauli principle.

We now proceed to perform a variational calculation for the two-electron system. Substituting (3-14) into (3-5), and integrating over the spin-coordinates, we obtain

$$\mathscr{H}\psi_j(1)\psi_j(2) = W_k\psi_j(1)\psi_j(2). \qquad (3\text{-}15)$$

Insertion of $\mathscr{H}$ from Eq. (3-1) into (3-15); integrating over the electronic coordinates, and making use of (3-13) leads to

$$\langle W_k \rangle = 2w_j + \iint |\psi_j(1)|^2 e^2/r_{12} |\psi_j(2)|^2 \, d\tau_1 \, d\tau_2. \qquad (3\text{-}16)$$

Using the antisymmetric product trial wavefunction (3-14), we obtain the total electronic energy of the system as a sum of the single orbital energies plus the electron-electron repulsion energy. We can write (3-16) as

$$\langle W_k \rangle = 2w_j + J_{j,j} \qquad (3\text{-}17)$$

with

$$J_{j,j} = \int\int |\psi_j(1)|^2 \, e^2/r_{12} \, |\psi_j(2)|^2 \, d\tau_1 \, d\tau_2. \qquad (3\text{-}18)$$

Let us now examine a case where two electrons occupy different space orbitals. The total electronic wavefunction, including the spin coordinates, has to be antisymmetric by exchange of the electronic coordinates.

The orbital function

$$\Psi_s = \sqrt{\tfrac{1}{2}} \, [\psi_i(1)\psi_j(2) + \psi_i(2)\psi_j(1)]$$

is symmetric, but

$$\Psi_a = \sqrt{\tfrac{1}{2}} \, [\psi_i(1)\psi_j(2) - \psi_i(2)\psi_j(1)] \qquad (3\text{-}20)$$

is antisymmetric.

For the spin wavefunctions, the products

$$\Phi_{s,1} = \alpha(1)\alpha(2), \qquad (3\text{-}21)$$

$$\Phi_{s,0} = \sqrt{\tfrac{1}{2}} \, [\alpha(1)\beta(1) + \alpha(2)\beta(1)], \qquad (3\text{-}22)$$

$$\Phi_{s,-1} = \beta(1)\beta(2) \qquad (3\text{-}23)$$

are symmetric, but

$$\Phi_{a,0} = \sqrt{\tfrac{1}{2}} \, [\alpha(1)\beta(2) - \alpha(2)\beta(1)] \qquad (3\text{-}24)$$

is antisymmetric.

The complete antisymmetric normalized wavefunctions, including both space and spin coordinates, are now obtained by multiplication:

$$(\text{sym.orb.}) \cdot (\text{antisym.spin}) = \Psi_s \Phi_{a,0}, \qquad (3\text{-}25)$$

$$(\text{antisym.orb.}) \cdot (\text{sym.spin}) = \Psi_a \cdot \left\{ \begin{array}{l} \Phi_{s,1} \\ \Phi_{s,0} \\ \Phi_{s,-1} \end{array} \right. \qquad (3\text{-}26)$$

By integration over the spin coordinates, we see that the four possible antisymmetric state functions are orthogonal to each other.

Let us find the energies of these states. We notice that, since the Hamlitonian operator is not a function of the spin, integration over the spin coordinates can immediately be carried out. The three wavefunctions that occur as $(\text{antisym.orb}) \cdot (\text{sym.spin})$, therefore, all have the same energy. Such a three-fold degenerate electronic state is called a *spin triplet.* Setting the spin degeneracy equal to $2S + 1$, we get $2S + 1 = 3$ or $S = 1$. On the other hand, the product $(\text{sym.orb.}) \cdot (\text{antisym.spin})$ is nondegenerate. We have $2S + 1 = 1$ or $S = 0$. Such a state is called a *spin singlet.* A molecule or atom in a state with $S > 0$ is *paramagnetic*; that is, it has a permanent magnetic moment. In all states for which $S = 0$, the atom or molecule is *diamagnetic.*

Defining the *Coulomb integral*

$$J_{i,j} = \int \int |\psi_i(1)|^2 \, \frac{e^2}{r_{12}} \, |\psi_j(2)|^2 \, d\tau_1 \, d\tau_2 \qquad (3\text{-}27)$$

and the *exchange integral*

$$K_{i,j} = \int \int \psi_i^*(1)\psi_j(1) \, \frac{e^2}{r_{12}} \, \psi_i(2)\psi_j^*(2) \, d\tau_1 \, d\tau_2 \qquad (3\text{-}28)$$

we find

$$\langle W \rangle_{S=1} = w_i + w_j + J_{i,j} - K_{i,j}, \qquad (3\text{-}29)$$

$$\langle W \rangle_{S=0} = w_i + w_j + J_{i,j} + K_{i,j}. \qquad (3\text{-}30)$$

Whenever product functions are employed as trial wavefunctions, the total electronic energy is given as a sum of orbital energies and electron-electron repulsion terms.

Notice that by demanding that the wavefunction be antisymmetric, we have introduced the "nonclassical" exchange integral into $\langle W \rangle$. A minimization of the energy when exchange is included is referred to as a *Hartree-Fock* calculation.

The J and K integrals are always positive, and taking $w_i + w_j$ to be the same for both states[1] we notice that the spin triplet is lower in energy than the spin singlet. The above is an example of Hund's first rule, which in general states: For a given electronic configuration the state with the highest value of S will have the lowest energy. Somewhat loosely, we may state that electrons prefer to occupy different space orbitals, since it is energetically favorable to avoid *spin pairing*.

Whether the ground state of a two-electron atom or molecule will be a spin singlet, corresponding to a $(\psi_j)^2$ configuration, or a spin triplet, corresponding to a $(\psi_j)^1(\psi_i)^1$ configuration, can be seen from Eq. (3-17) and Eq. (3-29). With $w_j < w_i$, we find that only provided $w_i - w_j > J_{i,i} - J_{i,j} + K_{i,j}$ will we have a spin singlet as the ground state; the loss of energy due to spin pairing is balanced against the orbital energy loss.

Only in a two-electron system is it possible to separate the orbital and spin functions. However, quite generally, we can write antisymmetric wavefunctions as determinants using spin orbitals. The spin-singlet wavefunction (3-14) can be written

---

[1] However, see Problem 2 at the end of this chapter.

$$\Psi(1,2) = (2!)^{-1/2} \begin{vmatrix} \overset{\alpha}{\psi_j}(1) & \overset{\alpha}{\psi_j}(2) \\ \overset{\beta}{\psi_j}(1) & \overset{\beta}{\psi_j}(2) \end{vmatrix} \quad . \tag{3-31}$$

For three electrons, a wavefunction may be

$$\Psi(1,2,3) = (3!)^{-1/2} \begin{vmatrix} \overset{\alpha}{\psi_j}(1) & \overset{\alpha}{\psi_j}(2) & \overset{\alpha}{\psi_j}(3) \\ \overset{\beta}{\psi_j}(1) & \overset{\beta}{\psi_j}(2) & \overset{\beta}{\psi_j}(3) \\ \overset{\alpha}{\psi_k}(1) & \overset{\alpha}{\psi_k}(2) & \overset{\alpha}{\psi_k}(3) \end{vmatrix} \quad . \tag{3-32}$$

We usually do not write a determinantal wavefunction in its entirety, but only give the diagonal term, writing, for instance, (3-32) as

$$\Psi(1,2,3) = |\overset{\alpha}{\psi_j}(1)\overset{\beta}{\psi_j}(2)\overset{\alpha}{\psi_k}(3)| . \tag{3-33}$$

If we exchange two columns in a determinant, the determinant changes sign. Thus, determinantal wavefunctions are antisymmetric. Furthermore, if the determinant has two identical rows, the determinant equals zero. In other words, determinantal wavefunctions have the Pauli principle built into them. It is important to realize that rigorous calculations of the physical properties of many-electron systems must be based on *antisymmetric state wavefunctions*, in which each and every electron coordinate is considered.

3.2 ATOMIC ELECTRONIC CONFIGURATIONS AND ENERGIES

We now turn to the general question of finding the ground states for atoms of the elements of the periodic table; these states are found by placing the electrons one by one into the lowest available atomic orbitals, with due account taken of both the Pauli principle and Hund's rule. Some examples are given in Table 3-1. The question of orbital degeneracy first arises for atomic carbon. Hund's rule states that the ground state for carbon may be $(1s)^2(2s)^2(2p_x)(2p_y)$, or $(1s)^2(2s)^2(2p_x)(2p_z)$, or $(1s)^2(2s)^2(2p_y)(2p_z)$, since these three configurations all can give spin triplets. We see that there are three *orbital configurations* for the two electrons and that these have the same energy. In other words, the ground state of carbon is three-fold spin degenerate and three-fold orbitally degenerate. The total degeneracy of the carbon ground state is therefore nine.

For nitrogen the ground state is given by the configuration $(1s)^2(2s)^2(2p_x)(2p_y)(2p_z)$, having a four-fold spin degeneracy. The ground state of nitrogen is not orbitally degenerate; the total degeneracy is therefore four. For oxygen the ground state configuration is $(1s)^2(2s)^2(2p_x)^2(2p_y)(2p_z)$, or one of the two other combinations. As for carbon the ground state of oxygen is totally nine-fold degenerate.

We shall now calculate the energy of a $(1s)^2$ configuration. Such a configuration corresponds to the ground state of helium (or helium-like systems). The one-electron Hamiltonian is

$$\mathscr{H}_1 = -(\hbar^2/2m)\nabla_1^2 - Qe^2/r_1 \qquad (3\text{-}34)$$

where for helium, $Q = 2$.

Table 3-1 Ground State Electronic Configurations for some Atoms.

Atom	Ground state electronic configuration
H	$(1s)$
He	$(1s)^2$
Li	$(1s)^2(2s)$
Be	$(1s)^2(2s)^2$
B	$(1s)^2(2s)^2(2p)$

We take a scaled $1s$ orbital,

$$(1s) = \sqrt{Z^3/\pi a_0^3} \exp(-Zr/a_0) \qquad (3\text{-}55)$$

with Z being a variational parameter.

The energy of the system is given by Eq. (3-17). Use of (3-34) and (3-35) yields

$$w(1s) = (Z^2/2 - QZ)e^2/a_0. \qquad (3\text{-}36)$$

Because

$$|(1s)(1)|^2 = Z^3/\pi a_0^3 \exp(-2Zr_1/a_0) \qquad (3\text{-}37)$$

is a spherical charge distribution, the potential that electron two experiences from electron one is, when $r_1 < r_2$, as if all the charge is collected at the center. On the other hand, when $r_1 > r_2$, it is constant, since electron two is then "inside" a charged sphere. Therefore

$$V(r_2) = \int_0^{r_2} e\,|(1s)(1)|^2\; \frac{1}{r_2}\; d\tau_1$$

$$+ \int_{r_2}^{\infty} e\,|(1s)(1)|^2\; \frac{1}{r_1}\; d\tau_1 \qquad (3\text{-}38)$$

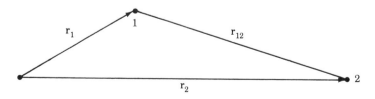

Figure 3-1   The potential felt by electron 2 under the influence of electron 1.

with

$$d\tau_1 = 4\pi r_1^2 dr_1. \tag{3-39}$$

By inserting the charge distribution given by (3-37) into (3-38) and integrating, we obtain

$$V(r_2) = \frac{e}{r_2} - \left(\frac{e}{r_2} + \frac{eZ}{a_0}\right) \exp\left(-\frac{2Z}{a_0} r_2\right). \tag{3-40}$$

The Coulomb integral is then calculated using (3-40):

$$J_{1s,1s} = \int_0^\infty \frac{Z^3}{\pi a_0^3} \exp\left(-\frac{2Z}{a_0} r_2\right) \left[\frac{e^2}{r_2} - \left(\frac{e^2}{r_2} + \frac{e^2 Z}{a_0}\right)\right.$$

$$\left. \times \exp\left(-\frac{2Z}{a_0} r_2\right)\right] 4\pi r_2^2\, dr_2 = \tfrac{5}{8} Z \left(\frac{e^2}{a_0}\right). \tag{3-41}$$

The total electronic energy of a $(1s)^2$ configuration is then, using (3-17), (3-36), and (3-41):

$$\langle W \rangle = (Z^2 - 2QZ + \tfrac{5}{8}Z) \frac{e^2}{a_0}. \tag{3-42}$$

Our example illustrates the Hartree-Fock method of calculating electronic structures of a many-electron system. Each electron experiences an averaged potential, obtained by integrating the charge densities of all the other electrons, and the one-particle Schrödinger equation is solved using this averaged potential.

Next, we minimize $\langle W \rangle$ with respect to the variational parameter Z,

$$\frac{d\langle W \rangle}{dZ} = 0 \quad \text{gives } Z_{\text{eff}} = Q - 5/16.$$

The minimized energy is therefore

$$W_{\min} = -(Q - 5/16)^2 e^2/a_0. \qquad (3\text{-}43)$$

For helium $Q = 2$, and the first ionization potential is calculated to be

$$\left[-\left(\frac{27}{16}\right)^2 + 2 \right] e^2/a_0 = 23.05 \text{ eV}. \qquad (\text{Exp: } 24.58 \text{ eV})$$

Detailed calculations show that for many-electron atomic systems the energy of an electron in an s orbital is different from that of an electron in one of the three-fold degenerate p orbitals in the same n quantum shell; this energy difference is due mainly to the different "effective nuclear charge" each type of electron experiences. Indeed, in a many-electron system, we know both from calculations and from experiment that the orbitals increase in energy according to $ns < np < nd$ (etc.), as shown in Figure 3-2.

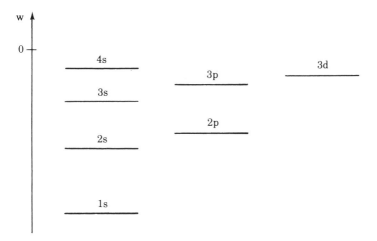

Figure 3-2 The one-electron energies for a many-electron atom. The drawing is only very approximate. The single s, p, and d levels are separated for the sake of clarity.

If the difference in energy between orbitals that are to receive electrons is small, as for example between the $4s$ and $3d$ orbitals, it is sometimes energetically favorable (refer to Hund's rule) to distribute the electrons in both orbitals. As an example, we find the lowest configurations for V, Cr, and Mn to be ([Ar] stands for a closed 18-electron shell)

$$V: \ [Ar](3d)^3(4s)^2,$$

$$Cr: \ [Ar](3d)^5(4s)^1,$$

$$Mn: \ [Ar](3d)^5(4s)^2.$$

Configurations such as $(3d)^5(4s)^1$ are, however, the exception rather than the rule. They occur for the first time in the transition elements.

Chapter 3

Suggested Reading

M. Karplus and R. N. Porter, Atoms and Molecules, Benjamin, New York, 1970.

J. C. Slater, *Phys. Rev., 34* (1929), 1293.
   The original treatment of determinantal wavefunctions.

A. Unsöld, *Ann. Physik, 82* (1927), 355.
   Calculation of the electronic states of He and He-like atoms.

C. Zener, *Phys. Rev., 36* (1930), 51; J. C. Slater, *ibid.*, 57.
   Calculation of atomic screening effects.

Problems

   1. Write the determinantal wavefunctions for the ground state of the oxygen atom, assuming that the unpaired electrons have α spin. What is the orbital degeneracy of the ground state? What is the spin degeneracy?

   2. Calculate the energies of the spin singlet and spin triplet states that arise from a $(1s)^1(2s)^1$ configuration. Is the sum of the single orbital minimized energies the same in the spin singlet and spin triplet? *Hint.* Use scaled orbitals,

$$(1s) = \sqrt{\frac{1}{\pi}} \left(\frac{Z}{a_0}\right)^{3/2} \exp\left(-\frac{Z}{a_0}r\right) \quad \text{and}$$

$$(2s) = \frac{1}{4\sqrt{2\pi}} \left(\frac{Z}{a_0}\right)^{3/2} \left(2 - \frac{Z}{a_0}r\right) \exp\left(-\frac{Z}{2a_0}r\right);$$

evaluate the kinetic and potential energies of the two orbitals using the Hamiltonian for a He-like system. Evaluate $J_{1s,2s}$ and $K_{1s,2s}$ as $J_{1s,1s}$ is evaluated in the text [the result: $J_{1s,2s} = Z(1/4 - 13/324)e^2/a_0$ and $K_{1s,2s} = Z(16/729)e^2/a_0$]. Minimize the total energy with respect to Z. Notice, however, that the state wavefunction for the excited spin singlet must be orthogonal to the ground spin singlet state wavefunction. This is only the case provided the screening factors are the same for $(1s)$ and $(2s)$. Having minimized the ground state energy and found $Z_{min} = Q - 5/16$, this is therefore the value we must choose also for the excited spin singlet state. But this is *not* the case for the excited spin triplet where we can vary Z freely.

The observed energy difference between the $(1s)^1(2s)^1$ spin triplet and the corresponding spin singlet is 0.80 eV for He, 1.74 eV for Li^+, and 3.06 eV for Be^{2+}.

ELECTRONIC STATES
OF MOLECULES

4-1 HYBRID ORBITALS

In a molecular orbital description of the bonding in molecules, atomic orbitals are combined to form LCAO-MO's. Electrons in such molecular orbitals are completely delocalized, that is, they can roam over the whole molecule. Molecular shapes are obtained from MO theory by minimizing the total electronic energy as a function of the interatomic distances and angles.

Another approach to molecular bonding starts by assigning each atom a certain number of "valences". Stereochemistry teaches us that these valences have directional properties. By choosing orbitals that have directional properties built into them, we can utilize structural information to construct bonding orbitals. These bonding orbitals are localized, as we have confined the electronic clouds to certain parts of the molecule.

In many cases it is convenient to use a mixture of localized and delocalized bonds to describe the electronic states of molecules. For this reason it is appropriate at this point to take up the matter of directed orbitals.

Directed orbitals are constructed from atomic orbitals. A linear combination of atomic orbitals located on one center is called a *hybrid orbital*. From an atomic set containing n orbitals, n linearly dependent hybrids can be constructed. Three types of directed orbitals play a particularly important role: the tetrahedral, the trigonal, and the linear hybrids.

Four equivalent tetrahedral hybrids can be constructed from an atomic set of s, p_x, p_y, and p_z orbitals. With a coordinate system as given in Figure 4-1, we notice that the linear combination of p orbitals that points from the center to corner one of the cube is given by

$$p_x + p_y + p_z.$$

Since we desire four equivalent hybrids, a fourth of the s orbital will have to be included. Normalizing to one, we obtain

$$\chi_{h1} = \tfrac{1}{2} (s + p_x + p_y + p_z). \qquad (4\text{-}1)$$

Similarly,

$$\chi_{h2} = \tfrac{1}{2} (s - p_x - p_y + p_z), \qquad (4\text{-}2)$$

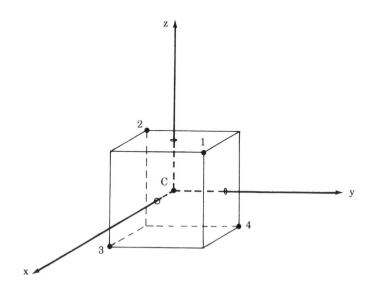

Figure 4–1 Coordinate system for tetrahedral hybridization.

$$\chi_{h3} = \tfrac{1}{2}(s + p_x - p_y - p_z), \tag{4-3}$$

$$\chi_{h4} = \tfrac{1}{2}(s - p_x + p_y - p_z). \tag{4-4}$$

These four linear combinations are called the sp^3 tetrahedral hybrid orbitals.

We can construct three orthonormal planar hybrids from the three atomic orbitals s, p_x, and p_y (see Figure 4-2). We exclude p_z, since that atomic orbital has a node in the xy plane. Directed from the center toward position 1, we take the trigonal hybrid

$$\chi_{h1} = c_1(p_x) + c_2(s). \tag{4-5}$$

No p_y contribution is used, since p_y has a node in the xz plane. Since we want to construct three equivalent orbitals, we must have $c_2^2 = \tfrac{1}{3}$. With $c_1^2 + c_2^2 = 1$, we obtain therefore

$$\chi_{h1} = \sqrt{\tfrac{1}{3}}(s) + \sqrt{\tfrac{2}{3}}(p_x). \tag{4-6}$$

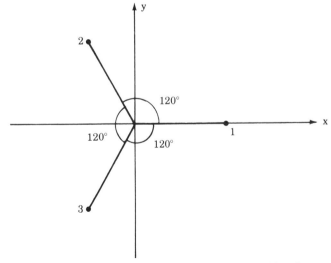

Figure 4-2   Coordinate system for trigonal hybridization.

The way we have placed the coordinate system, the remaining $\frac{1}{3}$ part of (p_x) and the whole of (p_y) must be divided equally between χ_{h2} and χ_{h3}. Therefore

$$\chi_{h2} = \sqrt{\tfrac{1}{3}}s - \sqrt{\tfrac{1}{6}}(p_x) + \sqrt{\tfrac{1}{2}}(p_y), \tag{4-7}$$

$$\chi_{h3} = \sqrt{\tfrac{1}{3}}s - \sqrt{\tfrac{1}{6}}(p_x) - \sqrt{\tfrac{1}{2}}(p_y). \tag{4-8}$$

χ_{1h}, χ_{2h}, and χ_{3h} are called the trigonal sp^2 hybrids.

Linear or sp hybrid orbitals are constructed as follows:

$$\chi_{h1} = \sqrt{\tfrac{1}{2}}(s + p_z), \tag{4-9}$$

$$\chi_{h2} = \sqrt{\tfrac{1}{2}}(s - p_z). \tag{4-10}$$

The contours of the linear hybrids are given in Figure 4-3.

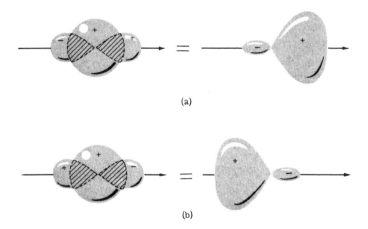

(a)

(b)

Figure 4–3 "Hybridization" of $2s$ and $2p$ orbitals. Contour surfaces of (a) $2s + 2p$, (b) $2s - 2p$. In the last case the minus sign is in front of the $2p$ orbital, effectively turning it around.

4-2   HOMONUCLEAR AND HETERONUCLEAR DIATOMIC MOLECULES

In a first approximation to the electronic structures of molecules, all the electrons the system possesses are placed in the molecular orbitals, taking due account of the Pauli principle. Although such an *electronic configuration* tells us which orbitals are occupied, and is useful in a qualitative sense, it does not provide information about the contributions to the energy from the omnipresent electron-electron repulsions. These inter-electronic-repulsion effects can be incorporated only by the use of *antisymmetric state wavefunctions*.

The molecular orbitals can be characterized by use of the relevant symmetry operators, and the same holds true for the electronic states. It is customary to denote the single molecular orbitals by lower case letters and the many-electronic wavefunctions by upper case letters.

The important symmetry operations for a linear molecule with an inversion center are $\hat{E}$, $\hat{C}_\phi$, $\hat{\sigma}_v$, and $\hat{i}$. Here $\hat{C}_\phi$ is a rotation of an arbitrary angle ϕ around the axis of the molecule. $\hat{\sigma}_v$ is a reflection in a plane containing that axis and $\hat{i}$ is the inversion operator. The characters for these symmetry operators are given in Table 4-1.

Table 4-1   Selected Characters for Linear Molecules with an Inversion Center.

	$\hat{E}$	$\hat{C}_\phi$	$\hat{\sigma}_v$	$\hat{i}$
σ_g^+	1	1	1	1
σ_u^+	1	1	1	-1
σ_g^-	1	1	-1	1
σ_u^-	1	1	-1	-1
π_g	2	$2\cos\phi$	0	2
π_u	2	$2\cos\phi$	0	-2
Δ_g	2	$2\cos 2\phi$	0	2
Δ_u	2	$2\cos 2\phi$	0	-2

Consider now the case of a homonuclear diatomic molecule. From the character table we notice that the one-fold degenerate orbitals that have eigenvalues of $+1$ under $\hat{C}_\phi$ are called sigma orbitals (this is consistent with the nomenclature used earlier). An eigenvalue of $+1$ for $\hat{\sigma}_v$ is indicated by a plus, whereas -1 is indicated by a minus sign. For a homonuclear diatomic molecule, for example, the molecular orbital $\psi = N(s_A + s_B)$ is a σ_g^+ orbital, whereas $\psi = N(s_A - s_B)$ is a σ_u^+ orbital.

The characters (Table 4-1) tell us further that for a doubly degenerate set of orbitals to be called π orbitals, they must have a character of $2 \cos \phi$ under $\hat{C}_\phi$. Consider now the two degenerate molecular orbitals

$$\psi_1 = N_1(p_x(A) + p_x(B)), \tag{4-11}$$

$$\psi_2 = N_1(p_y(A) + p_y(B)) \tag{4-12}$$

pictured in Figure 4-4. We notice the way the coordinate systems have been placed and get therefore for a rotation by $\hat{C}_\phi$ around the A–B axis

$$\hat{C}_\phi \begin{pmatrix} x \\ y \end{pmatrix} = \begin{pmatrix} \cos \phi & -\sin \phi \\ \sin \phi & \cos \phi \end{pmatrix} \begin{pmatrix} x \\ y \end{pmatrix}. \tag{4-13}$$

Because p_x and p_y behave as x and y, we find

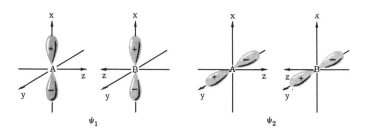

Figure 4-4   Coordinate systems and p orbitals for a diatomic molecule.

$$\hat{C}_{\phi}\psi_1 = (\cos \phi)\,\psi_1 - (\sin \phi)\,\psi_2, \qquad (4\text{-}14)$$

$$\hat{C}_{\phi}\psi_2 = (\sin \phi)\,\psi_1 + (\cos \phi)\,\psi_2. \qquad (4\text{-}15)$$

Written as a matrix Eqs. (4-14) and (4-15) can be combined

$$\hat{C}_{\phi}\begin{pmatrix}\psi_1 \\ \psi_2\end{pmatrix} = \begin{pmatrix}\cos \phi & -\sin \phi \\ \sin \phi & \cos \phi\end{pmatrix}\begin{pmatrix}\psi_1 \\ \psi_2\end{pmatrix}.$$

The trace of this transformation matrix is $2 \cos \phi$. The character of ψ_1 and ψ_2 under $\hat{C}_{\phi}$ is therefore that of a pi-orbital set.

Furthermore we see that

$$\hat{i}\begin{pmatrix}\psi_1 \\ \psi_2\end{pmatrix} = \begin{pmatrix}-1 & 0 \\ 0 & -1\end{pmatrix}\begin{pmatrix}\psi_1 \\ \psi_2\end{pmatrix} \qquad (4\text{-}16)$$

with trace -2. Comparison with Table 4-1 therefore tells us that the "bonding" set of orbitals given by ψ_1 and ψ_2 in Eqs. (4-11) and (4-12) transforms like a π_u set of orbitals. The "antibonding" set

$$\psi_3 = N_2(p_x(A) - p_x(B)), \qquad (4\text{-}17)$$

$$\psi_4 = N_2(p_y(A) - p_y(B)) \qquad (4\text{-}18)$$

would on the other hand transform like a π_g set of orbitals. Schematic molecular orbital diagrams for homonuclear diatomic molecules are shown in Figure 4-5.

The results of exact calculations of the energy levels of H_2^+ as a function of the internuclear distance R are diagrammed in Figure 4-6. In view of what has been said about noncrossing rules for levels having the same symmetry, we should not expect $2\sigma_g^+$ and $3\sigma_g^+$, e.g., to cross. The reason that they do is that a linear molecule possesses symmetry properties that we have not discussed here. When this "extra symmetry" is taken into account, the two σ_g^+ orbitals no longer transform identically; thus, they are allowed to cross.

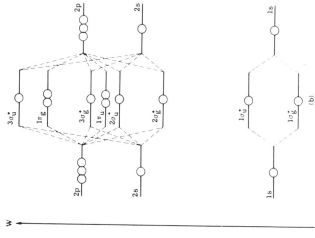

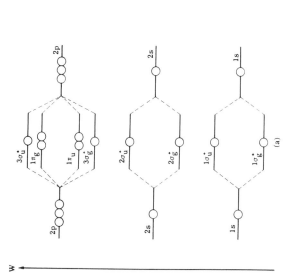

Figure 4-5   Energy diagrams for homonuclear diatomic molecules. The energy level designations give the symmetries of the corresponding wavefunctions. The numbering begins at the lowest energy level and goes up, each symmetry species is numbered separately. Diagram (a) is for no 2s-2p mixing while (b) represents substantial 2s-2p mixing.

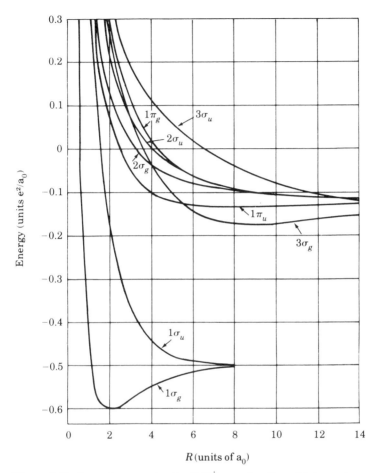

Figure 4-6 Lowest energy levels of H_2^+ as a function of internuclear distance with internuclear repulsion included. Adapted from J. C. Slater, Quantum Theory of Matter, 2nd ed., McGraw-Hill, New York, 1968, p. 376.

For the hydrogen molecule the lowest electronic configuration has two electrons in the $(1\sigma_g^+)$ molecular orbitals. A configuration $(1\sigma_g^+)^2$ gives rise to one state with the wavefunction

$$\psi = |(1\bar{\sigma}_g^+)(1\bar{\sigma}_g^+)| \tag{4-19}$$

or written out

$$\psi = \sqrt{\tfrac{1}{2}}[(1\bar{\sigma}_g^+)(1)(1\bar{\sigma}_g^+)(2)-(1\bar{\sigma}_g^+)(1)(1\bar{\sigma}_g^+)(2)]. \tag{4-20}$$

Because we have only two electrons, we can separate the space functions from the spin functions and write (4-20) as

$$\psi = (1\sigma_g^+)(1)(1\sigma_g^+)(2)\sqrt{\tfrac{1}{2}}[\alpha(1)\beta(2) - \beta(1)\alpha(2)]. \tag{4-21}$$

The space function $(1\sigma_g^+)(1)(1\sigma_g^+)(2)$ is easily seen to transform like a Σ_g^+ function. The spin function is one-fold degenerate, and for the spin multiplicity we have $2S + 1 = 1$; $S = 0$. The name of the ground state of the hydrogen molecule is "singlet-sigma-g-plus" written $^1\Sigma_g^+$, the spin multiplicity being given as an upper left superscript.

Low-lying excited electronic states for the hydrogen molecule are obtained by promoting one of the $(1\sigma_g^+)$ electrons to the $(1\sigma_u^+)$ orbital (Figure 4-5). Since the two orbitals are different in the configuration $(1\sigma_g^+)^1(1\sigma_u^+)^1$, we obtain both a spin triplet and a spin singlet. For the orbital part, we get $\hat{E}\Psi = 1\Psi$, $\hat{C}_\varphi\Psi = 1\Psi$, $\hat{\sigma}_v\Psi = 1\Psi$, $\hat{i}\Psi = -1\Psi$. The configuration $(1\sigma_g^+)^1(1\sigma_u^+)^1$ gives rise to two states: a "triplet-sigma-u-plus", $^3\Sigma_u^+$, and a "singlet-sigma-u-plus", $^1\Sigma_u^+$. According to Hund's first rule, the $^3\Sigma_u^+$ state has lower energy than the $^1\Sigma_u^+$ state.

The next excited configuration occurs if both the $(1\sigma_g^+)$ electrons are excited into the $(1\sigma_u^+)$ orbital. The configuration $(1\sigma_u^+)^2$ gives rise to a $^1\Sigma_g^+$ state. We notice that, in general, filled electron "shells" give a state that is totally symmetric under the symmetry operations. This is due to the fact that there are no "degrees of freedom" in such a case.

The calculation of the energy of the ground state of H_2 is carried out as follows. As variational function we take the state wavefunction (4-19) in which

$$(1\sigma_g^+) = \frac{1}{\sqrt{2 + 2S}} \; (\chi_a + \chi_b) \qquad (4\text{-}22)$$

where χ_a stands for a scaled $(1s)$ atomic orbital $\sqrt{Z^3/\pi a_0^3}$ $\exp(-Zr_a/a_0)$ centered on proton A and χ_b for a similar function centered on proton B. By integrating over the spin-coordinates, we obtain

$$\langle W \rangle = \int\!\!\int (1\sigma_g^+)(1)(1\sigma_g^+)(2)\,\mathscr{H}$$
$$\times (1\sigma_g^+)(1)(1\sigma_g^+)(2)\,d\tau_1\,d\tau_2 \qquad (4\text{-}23)$$

with

$$\mathscr{H} = -\frac{\hbar^2}{2m}\nabla_1^2 \; - \; \frac{\hbar^2}{2m}\nabla_2^2 \; + \; V_1 \; + \; V_2 \; + \; \frac{e^2}{r_{12}}$$
$$(4\text{-}24)$$

Substituting (4-24) into (4-23) we get

$$\langle W \rangle = 2w(1\sigma_g^+)$$
$$+ \int\!\!\int [(1\sigma_g^+)(1)]^2 e^2/r_{12} \; [(1\sigma_g^+)(2)]^2 \; d\tau_1\,d\tau_2. \qquad (4\text{-}25)$$

We calculate $\langle W \rangle$ as a function of the interatomic distance R and the scaling parameter Z. Minimizing with respect to these two parameters, we get $R_{eq} = 1.38a_0$ and $Z_{eff} = 1.20$. The binding energy is 3.47 eV. The experimental values are $R_{eq} = 1.40a_0$ and $D_e = 4.72$ eV.

A plot of the electronic energy with internuclear repulsion included as a function of the interatomic distance is called the *potential function* for a state of the molecule. The potential functions for the ground and lowest excited states of H_2 are shown in Figure 4-7.

If we expand the space part of the wavefunction for the ground state of the hydrogen molecule, we obtain

$$\sigma_g^+(1)\sigma_g^+(2) = \chi_a(1)\chi_b(2) + \chi_a(2)\chi_b(1) + \chi_a(1)x_a(2)$$
$$+ \chi_b(1)\chi_b(2). \qquad (4\text{-}26)$$

The last two terms of the expanded function correspond to an electronic density distribution in which both electrons are as-

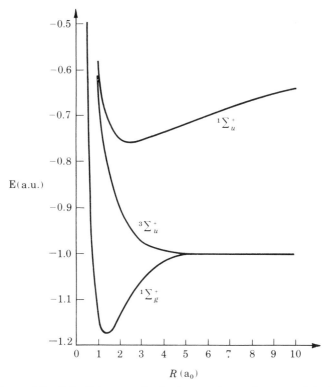

Figure 4-7   Potential curves for the lowest electronic states of the H_2 molecule.

sociated with the same hydrogen nucleus. Therefore, they correspond to the electronic "ionic" configurations, H^-H^+ and H^+H^-. They appear with the same weight in the wavefunction as the "covalent" structure represented in the first two terms. Looking at the wavefunction (4-26) as a trial function, we may indeed omit the last two terms and take

$$\Psi = \chi_a(1)\chi_b(2) + \chi_a(2)\chi_b(1). \qquad (4\text{-}27)$$

This trial wavefunction is called a *Heitler-London wavefunction*, or a *valence bond wavefunction*. For the ground state of hydrogen, using a scaled $(1s)$ atomic function, the equilibrium distance R_{eq} is found to be $1.43a_0$, with a dissociation energy of 3.76 eV. Z_{eff} is found to be 1.17.

Finally, we may consider both "ionic" and "covalent" terms in the wave function, giving these terms different weight. This corresponds to taking a variational wavefunction

$$\Psi = N[\chi_a(1)\chi_b(2) + \chi_a(2)\chi_b(1) + \mu[\chi_a(1)\chi_a(2) + \chi_b(1)\chi_b(2)]] \qquad (4\text{-}28)$$

where N is a normalizing constant and μ a variational parameter. Using scaled $(1s)$ functions with $Z_{eff} = 1.19$, the dissociation energy is calculated to be 4.10 eV. At the equilibrium distance $\mu \simeq 0.25$; the ionic structure contributes 6% to the total wavefunction.

Consider next the homonuclear diatomic electronic configuration $(1\sigma_g^+)^2(1\sigma_u^+)^2$. We notice from the expression for the orbital energies (Eq. 2-20) that $(1\sigma_u^+)$ is placed higher above H_{AA} than $(1\sigma_g^+)$ is placed below. Thus an input of energy is required to form a molecule with such an electronic configuration. For example, two isolated He atoms have lower energy than a hypothetical helium molecule; this is in agreement with the fact that He_2 does not exist in nature.

With

$$(1\sigma_g^+) = \frac{1}{\sqrt{2+2S}} ((1s)_A + (1s)_B), \qquad (4\text{-}29)$$

$$(1\sigma_u^+) = \frac{1}{\sqrt{2 - 2S}} ((1s)_A - (1s)_B) \qquad (4\text{-}30)$$

we observe that the charge distribution for the $(1\sigma_g^+)^2(1\sigma_u^+)^2$ configuration is obtained by summing the contributions of all four electrons:

$$2\left(\frac{1}{2 + 2S}\right)((1s)_A + (1s)_B)^2 + \left(\frac{1}{2 - 2S}\right)((1s)_A - (1s)_B)^2$$

$$= \frac{2}{1 - S^2} ((1s)^2_A + (1s)^2_B - 2S\,(1s)_A(1s)_B).$$
$$(4\text{-}31)$$

For small values of S the charge distribution is approximately $2(1s)_A^2 + 2(1s)_B^2$. This is just the value for the case in which there is no "bonding" between the atoms. In general, the overlap between two $(1s)$ orbitals in diatomic molecules in which both atoms have $Q > 2$ is very small: the high nuclear charge draws the $(1s)$ electrons close to the nucleus. Consequently, we can neglect the bonding between $(1s)$ orbitals in such molecules.

Using the molecular orbital diagrams in Figure 4-5 we find that other representative diatomic molecules have the following ground electronic configurations and states:

$$Li_2 \quad (1s)_A^2(1s)_B^2(2\sigma_g^+)^2,\ {}^1\Sigma_g^+.$$

$$N_2 \quad (1s)_A^2(1s)_B^2(2\sigma_g^+)^2(2\sigma_u^+)^2(2\sigma_g^+)^2(1\pi_u)^4,\ {}^1\Sigma_g^+.$$

The ground state electronic configuration for O_2 is

$$(1s)_A^2(1s)_B^2(2\sigma_g^+)^2(2\sigma_u^+)^2(3\sigma_g^+)^2(1\pi_u)^4(1\pi_g)^2.$$

Since $(1\pi_g)$ can accommodate a total of four electrons, the two electrons of O_2 that go into $(1\pi_g)$ will be able to occupy different orbitals, $(\pi_g^x)(\pi_g^y)$. Such a configuration can give rise to a spin triplet state ($S = 1$) as well as a spin singlet ($S = 0$) state. Hund's rule then tells us that the spin-triplet state has the lower energy. The ground state of oxygen therefore shows paramagnetism.

The wavefunctions for the triplet are

$$\frac{1}{\sqrt{2}} \left[\pi_g^x(1)\pi_g^y(2) - \pi_g^x(2)\pi_g^y(1) \right] \begin{cases} \alpha(1)\alpha(2), \\[2mm] \dfrac{1}{\sqrt{2}} \left(\alpha(1)\beta(2) + \alpha(2)\beta(1) \right), \\[2mm] \beta(1)\beta(2). \end{cases} \quad (4\text{-}32)$$

With

$$\hat{C}_\varphi(\pi_g^x) = (\cos \varphi)(\pi_g^x) + (\sin \varphi)(\pi_g^y), \quad (4\text{-}33)$$

$$\hat{C}_\varphi(\pi_g^y) = -(\sin \varphi)(\pi_g^x) + (\cos \varphi)(\pi_g^y), \quad (4\text{-}34)$$

we find for the space part of the wavefunctions (4-32)

$$\Psi = \frac{1}{\sqrt{2}} \left[\pi_g^x(1)\pi_g^y(2) - \pi_g^x(2)\pi_g^y(1) \right] \quad (4\text{-}35)$$

that $\hat{C}_\varphi \Psi = 1\Psi$, $\hat{\sigma}_v \Psi = -1\Psi$ and $\hat{i}\Psi = 1\Psi$. The ground state of oxygen is therefore $^3\Sigma_g^-$.

For a diatomic molecule AB in which the atomic nuclei are different the LCAO-MO functions are still given by $\psi = c_1\chi_A + c_2\chi_B$. However, in this case, symmetry does not demand $|c_1| = |c_2|$, and we write

$$\psi = N(\chi_A + \lambda\chi_B). \quad (4\text{-}36)$$

The constant λ determines the "polarity" of the orbital—the larger the value of λ the greater the chance of finding the electron around nucleus B. The normalization constant N is given by the condition

$$N^2(1 + \lambda^2 + 2\lambda S) = 1 \quad (4\text{-}37)$$

with

$$S = \int \chi_A^* \chi_B \, d\tau \qquad (4\text{-}38)$$

Let us assume that nucleus A has a positive charge of $+Q_A|e|$ and nucleus B a positive charge $+Q_B|e|$, with $Q_B > Q_A$. The orbital energies of the atomic orbitals centered on B will then have lower energies than the corresponding orbitals centered on A. The molecular orbitals that can be formed using $1s(A)$ and $1s(B)$ are illustrated in Figure 4-8.

To construct an orbital energy diagram for a diatomic molecule in which the two nuclei are different, we must remember that orbitals with the same symmetry repel each other. For diatomic molecules in which the two nuclei are not far from each other in the periodic table, the diagram will be approximately as shown in Figure 4-9.

We now place electrons one after the other into the energy diagram (Figure 4-9). For example, the electronic ground state configuration of CO ($6 + 8 = 14$ electrons) is $(1\sigma^+)^2(2\sigma^+)^2(3\sigma^+)^2(1\pi)^4(4\sigma^+)^2(5\sigma^+)^2$, $^1\Sigma^+$. The electronic ground state con-

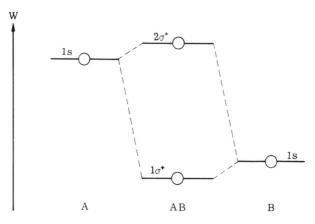

Figure 4-8 A combination of $1s$ orbitals with different energies.

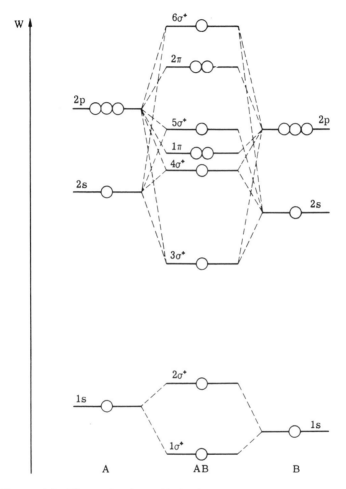

Figure 4-9 The molecular orbitals for a diatomic molecule with different atomic nuclei. The diagram indicates that $4\sigma^+$ repels $5\sigma^+$, making $5\sigma^+$ less stable than the 1π orbitals.

figuration of NO is $(1\sigma^+)^2(2\sigma^+)^2(3\sigma^+)^2(1\pi)^4(4\sigma^+)^2(5\sigma^+)^2 \cdot$
$(2\pi)^1$, $^2\Pi$. The last electron goes into an antibonding π orbital.
The molecule is paramagnetic since it possesses an "unpaired"
electron.

On the other hand, if the two nuclei have different charges, as
for example the case of LiH, it is a good approximation to regard
the Li $(1s)$ orbital as "nonbonding" and suppose that the chemical
bonding between Li and H takes place between the H $(1s)$ and a
linear hybrid of $(2s)$ and $(2p_z)$ lithium atomic orbitals. The
electronic ground state configuration of LiH is then $(1s)^2_{Li}(\sigma^+)^2$,
$^1\Sigma^+$, where $(\sigma^+) = N((\chi_h)_{Li} + \lambda(1s)_H)$.

Consider now the bonding in NaCl. The atomic ground state
electronic configurations of Na and Cl are respectively

$$\text{Na} \quad \ldots(2p)^6(3s)^1,$$

$$\text{Cl} \quad \ldots(2p)^6(3s)^2(3p)^5.$$

To account for the bonding Na-Cl we form linear hybrids of the
$(3s)$ and $(3p_z)$ atomic orbitals on both Na and Cl, and, by
pointing the z axes of the Na and Cl coordinate systems toward
each other, we obtain

$$\chi^+_{hCl} = \frac{1}{\sqrt{2}}[(3s)_{Cl} + (3p_z)_{Cl}], \qquad (4\text{-}39)$$

$$\chi^-_{hCl} = \frac{1}{\sqrt{2}}[(3s)_{Cl} - (3p_z)_{Cl}], \qquad (4\text{-}40)$$

$$\chi^+_{hNa} = \frac{1}{\sqrt{2}}[(3s)_{Na} + (3p_z)_{Na}], \qquad (4\text{-}41)$$

$$\chi^-_{hNa} = \frac{1}{\sqrt{2}}[(3s)_{Na} - (3p_z)_{Na}]. \qquad (4\text{-}42)$$

Let the bonding sigma molecular orbitals in Na-Cl be given by

$$\psi_\sigma = N(\chi^+_{hCl} + \lambda\chi^+_{hNa}). \qquad (4\text{-}43)$$

The ground state electronic configuration of NaCl is then

$$\ldots (2p)_{Na}^{6} (2p)_{Cl}^{6} (3p)_{Cl}^{4} (\chi_{hCl}^{-})^2 (\psi_{\sigma})^2.$$

Neglecting spin, and concentrating only on the bonding ψ_{σ} molecular orbital, the orbital wavefunction for the ground state of NaCl is

$$\Psi = N^2[\chi_{hCl}^{+}(1) + \lambda\chi_{hNa}^{+}(1)][\chi_{hCl}^{+}(2) + \lambda\chi_{hNa}^{+}(2)]$$
$$(4\text{-}44)$$

or, in expanded form,

$$\Psi = N^2[\chi_{hCl}^{+}(1)\,\chi_{hCl}^{+}(2) + \lambda^2\chi_{hNa}^{+}(1)\,\chi_{hNa}^{+}(2)\}$$
$$+ N^2\lambda\{\chi_{hNa}^{+}(1)\,\chi_{hCl}^{+}(2) + \chi_{hCl}^{+}(1)\,\chi_{hNa}^{+}(2)\}.$$
$$(4\text{-}45)$$

The first set of parentheses gives the part of the wavefunction corresponding to an "ionic" electronic configuration, the second to a "covalent" structure. We can now calculate the bonding energy of NaCl using either the complete molecular orbital wavefunction (4-45), or we can investigate what the "ionic" and the "covalent" parts of the wavefunction will yield when taken by themselves. The potential curves for the three calculations are given in Figure 4-10. Notice that at the bond equilibrium distance, $R_{eq} = 4.45a_0$, the energy is to a very good approximation given by assuming an "ionic" wavefunction. However, at $R = 18a_0$ the ground state becomes primarily a "covalent" state, corresponding to the fact that the dissociation products of NaCl are Na and Cl, and not Na^+ and Cl^-.

4-3 WATER

In its electronic ground state H_2O has an angular configuration and we take the H-O-H angle to be 2θ. We shall now give a simplified molecular orbital treatment of the system. Let us place the water molecule in a coordinate system as given in

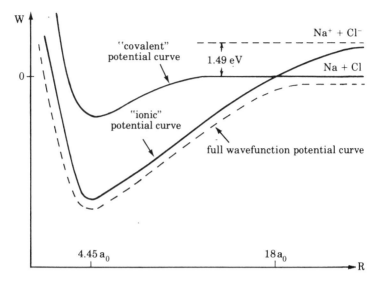

Figure 4-10   Potential energy curves for NaCl.

Figure 4-11. Among other symmetry operations, the molecule has a $\hat{\sigma}_v$ symmetry operation, namely, reflection in the xz plane. Placing a $(1s)$ orbital on each hydrogen nucleus and putting the overlap integral between the two hydrogen orbitals equal to zero, we form normalized linear combinations of $(1s)_A$ and $(1s)_B$ that are eigenfunctions of $\hat{\sigma}_v$:

With

$$\psi_g = \frac{1}{\sqrt{2}}((1s)_A + (1s)_B) \qquad (4\text{-}46)$$

we have

$$\hat{\sigma}_v \psi_g = \psi_g \qquad (4\text{-}47)$$

and with

$$\psi_u = \frac{1}{\sqrt{2}}((1s_A) - (1s_B)) \qquad (4\text{-}48)$$

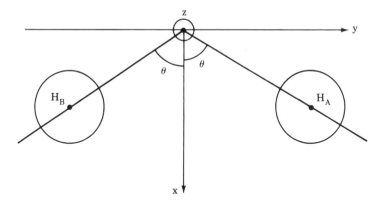

Figure 4-11   Coordinate systems for the calculation of the molecular orbital energy of H_2O.

we get

$$\hat{\sigma}_v \psi_u = - \psi_u. \tag{4-49}$$

On the oxygen atom we take the $(2s)$ and $(2p_x)$, $(2p_y)$, $(2p_z)$ atomic orbitals as "valence orbitals". The symmetry operations in H_2O are: $\hat{E}$, $\hat{C}_2(x)$, $\hat{\sigma}_v(xy)$, and $\hat{\sigma}_v(xz)$. We now take the valence orbitals and find the eigenvalues of the symmetry operators. We can then differentiate between the orbitals according to their transformation behavior

	$\hat{E}$	$\hat{C}_2(x)$	$\hat{\sigma}_v(xy)$	$\hat{\sigma}(xz)$	Designation
$(s)_0$	1	1	1	1	a_1
$(p_x)_0$	1	1	1	1	a_1
$(p_y)_0$	1	-1	1	-1	b_1
$(p_z)_0$	1	-1	-1	1	b_2
$(\psi_g)_H$	1	1	1	1	a_1
$(\psi_u)_H$	1	-1	1	-1	b_1

Only orbitals which transform the same way can be combined together in a symmetry-adapted molecular orbital. We have therefore

$$\psi(a_1) = c_1\psi_g + c_2 p_x + c_3(s),\qquad(4\text{-}50)$$

$$\psi(b_1) = c_4\psi_u + c_5 p_y,\qquad(4\text{-}51)$$

$$\psi(b_2) = p_z.\qquad(4\text{-}52)$$

First of all, we notice that the $(p_z)_O$ orbital does not participate in the bonding. Secondly, in order to focus on the principal bonding features in water, we shall put $c_3 = 0$, thereby not including $(s)_O$ as a valence orbital. If we further put all overlap integrals equal to zero, we obtain the two secular equations that determine the orbital energies of $\psi(a_1)$ and $\psi(b_1)$

$$a_1: \qquad \begin{vmatrix} H_s - w & H_g \\[2mm] H_g & H_p - w \end{vmatrix} = 0,\qquad(4\text{-}53)$$

$$b_1: \qquad \begin{vmatrix} H_s - w & H_u \\[2mm] H_u & H_p - w \end{vmatrix} = 0,\qquad(4\text{-}54)$$

where

$$H_s = \int \psi_g \mathscr{H} \psi_g \, d\tau = \int \psi_u \mathscr{H} \psi_u \, d\tau,\qquad(4\text{-}55)$$

$$H_p = \int p_x \mathscr{H} p_x \, d\tau = \int p_y \mathscr{H} p_y \, d\tau,\qquad(4\text{-}56)$$

$$H_g = \int \psi_g \mathscr{H} p_x \, d\tau\qquad(4\text{-}57)$$

and

$$H_u = \int \psi_u \mathscr{H} p_y \, d\tau.\qquad(4\text{-}58)$$

$\mathscr{H}$ is the Hamiltonian for the system.

The lowest root of each of the two secular equations is

$$w' = \frac{H_s + H_p}{2} - \tfrac{1}{2} \sqrt{(H_s - H_p)^2 - 4H_g^2}, \quad (4\text{-}59)$$

$$w'' = \frac{H_s + H_p}{2} - \tfrac{1}{2} \sqrt{(H_s - H_p)^2 - 4H_u^2}. \quad (4\text{-}60)$$

The two corresponding molecular orbitals are called $\psi^b(a_1)$ and $\psi^b(b_1)$, respectively. As there are ten electrons in the water molecule, the electronic ground state configuration of the molecule is therefore $(1s)_0^2 (2s)_0^2 (2p_z)_0^2 (\psi^b(a_1))^2 (\psi^b(b_1))^2$. The state designation is $^1A_1$.

Neglecting the e^2/r_{12} contributions to the total energy and concentrating on the four electrons in $\psi^b(a_1)$ and $\psi^b(b_1)$, we obtain the molecular electronic energy of the ground state

$$W = W_0 + 2(H_s + H_p) - \sqrt{(H_s - H_p)^2 - 4H_g^2}$$
$$- \sqrt{(H_s - H_p)^2 - 4H_u^2}. \quad (4\text{-}61)$$

We observe that the molecular integrals that occur in this expression are functions of the bond angle 2θ and the bond distance $O - H$ (Figure 4-11). If we minimize W with respect to θ, the functional dependence occurs only through H_g and H_u, since we have placed all overlap integrals equal to zero. By expansion, we see

$$H_g = \frac{1}{\sqrt{2}} \int (1s)_A \mathcal{H} p_x \, d\tau + \frac{1}{\sqrt{2}} \int (1s)_B \mathcal{H} p_x \, d\tau. \quad (4\text{-}62)$$

We can resolve p_x into $\cos \theta$ times a p_σ orbital directed along the $O - H_A$ bond and $\sin \theta$ times a p_π orbital directed perpendicular to the $O - H_A$ bond. Therefore, with

$$\beta_{OH} = \int p_\sigma \mathcal{H} s_H \, d\tau \qquad (4\text{-}63)$$

$$\int (1s)_A \mathcal{H} p_x \, d\tau = \int (1s)_A \mathcal{H} (p_\sigma \cos \theta + p_\pi \sin \theta) \, d\tau$$

$$= \beta_{OH} \cos \theta \qquad (4\text{-}64)$$

since the $\sin \theta$ term vanishes on symmetry grounds. The same expression is obtained for $\int (1s)_B \mathcal{H} p_x \, d\tau$ and we find

$$H_g = \sqrt{2} \, \beta_{OH} \cos \theta. \qquad (4\text{-}65)$$

Similarly

$$H_u = \sqrt{2} \, \beta_{OH} \sin \theta. \qquad (4\text{-}66)$$

Substituting (4-65) and (4-66) into (4-61) leads to

$$W = W_0 + 2(H_s + H_p) - \sqrt{(H_s - H_p)^2 - 8\beta_{OH}^2 \cos^2 \theta} \\ - \sqrt{(H_s - H_p)^2 - 8\beta_{OH}^2 \sin^2 \theta} \qquad (4\text{-}67)$$

Evaluating $\partial W/\partial \theta$ and putting it equal to zero yields the equation for the extremum points

$$\left[\frac{1}{\sqrt{(H_s - H_p)^2 - 8\beta_{OH}^2 \cos^2 \theta}} \right. \\ \left. - \frac{1}{\sqrt{(H_s - H_p)^2 - 8\beta_{OH}^2 \sin^2 \theta}} \right] \cos \theta \sin \theta = 0. \qquad (4\text{-}68)$$

The solutions $\theta = n\pi/2$, $n = 0, 1, 2, \ldots$ lead to maxima but $\cos^2 \theta = \sin^2 \theta$ or $\theta = \pi/4$ produces a minimum for W. Our simple treatment, therefore, predicts a bond angle $2\theta = 90°$ in H-O-H.

Looking again at the set of molecular orbitals (4-50, 4-51, 4-52) we notice that the linear combinations

$$\phi_1 = \psi(a_1) + \psi(b_1) + \psi(b_2), \qquad (4\text{-}69)$$

$$\phi_2 = \psi(a_1) - \psi(b_1) - \psi(b_2) \qquad (4\text{-}70)$$

with $c_1 = c_4$ and $c_2 = c_3 = c_5 = 1$ will give tetrahedral valence hybrid orbitals directed from the oxygen atom toward the two hydrogen atoms. Inclusion of the $(2s)$ orbital and fixing the coefficients will evidently make the H-O-H angle equal to the tetrahedral angle of $109°28'$. We can expect, therefore, that the actual H-O-H angle will be somewhere in between $90°$ and $109°28'$. Experiments show it to be about $105°$. Thus the $(2s)$ orbital will have to be included among the valence orbitals of water.

Had we at the outset used the tetrahedral hybrids on the oxygen atom, two of the sp^3 hybrids could have been used to form bonds with the two hydrogen atoms, and the remaining two hybrids would have contained nonbonding "lone pairs" of electrons. The electronic ground state of water in this formulation is

$$^1A_1 \; (\mathrm{h}_1^b)^2 \; (\mathrm{h}_2^b)^2 \; (\mathrm{h}_3)^2 (\mathrm{h}_4)^2.$$

This is called a *localized* description, because we consider one O-H bond as independent of the other one. It is, however, a *delocalized* set of orbitals (4-50, 4-51, 4-52) that provides the most general description of the bonding in the water molecule. To convert to the localized description, we have to place restrictions on some of the variational parameters, leading to a less flexible description. However, we have the advantage that in a localized description we can use our chemical intuition to pinpoint the location of the electronic densities in a molecule.

4-4 CARBON DIOXIDE

The CO_2 molecule is known from electron diffraction measurements to be linear in its ground state. The bonding in the molecule is taken care of by sixteen valence electrons—the six $(2s)^2(2p)^4$ electrons from one oxygen atom and the four $(2s)^2(2p)^2$ from the carbon atom. For a linear molecule possessing a center of inversion, we characterize the orbitals according to the symmetry operations $\hat{C}_\phi$, $\hat{\sigma}_v$, and $\hat{i}$ (Figure 4-12). For the carbon orbitals, we obtain

$$2s(C) \qquad \sigma_g^+$$
$$2p_z(C) \qquad \sigma_u^+$$
$$\left.\begin{array}{l} 2p_x(C) \\[1em] 2p_y(C) \end{array}\right\} \pi_u$$

For the oxygen orbitals it is practical first to form linear combinations, and then to find the transformation properties of these combinations:

$$\frac{1}{\sqrt{2}}\ (2s_1 + 2s_2) \quad \sigma_g^+$$
$$\frac{1}{\sqrt{2}}\ (2s_1 - 2s_2) \quad \sigma_u^+$$

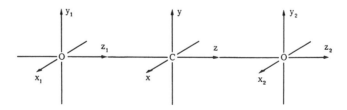

Figure 4-12 Coordinate system for the ground state of the CO_2 molecule.

$$\frac{1}{\sqrt{2}} \, (1p_{z_1} + 2p_{z_2}) \quad \sigma_u^+$$

$$\frac{1}{\sqrt{2}} \, (2p_{z_1} - 2p_{z_2}) \quad \sigma_g^+$$

$$\left. \begin{array}{l} \dfrac{1}{\sqrt{2}} \, (2p_{x_1} + 2p_{x_2}) \qquad \pi_u^a \\[2ex] \dfrac{1}{\sqrt{2}} \, (2p_{y_1} + 2p_{y_2}) \qquad \pi_u^b \end{array} \right.$$

$$\left. \begin{array}{l} \dfrac{1}{\sqrt{2}} \, (2p_{x_1} - 2p_{x_2}) \qquad \pi_g^a \\[2ex] \dfrac{1}{\sqrt{2}} \, (2p_{y_1} - 2p_{y_2}) \qquad \pi_g^b \end{array} \right.$$

We now combine the oxygen orbitals with the appropriate carbon orbitals to obtain the molecular orbitals for CO_2. Remember that only orbitals that transform in the same way can be combined. We have, for example, the π_u orbitals $\psi(\pi_u) = \alpha \psi C(\pi_u) + \beta \psi [O \cdots O](\pi_u)$. The coefficients α and β have the same sign in the bonding orbital and different signs in the antibonding orbital. The final σ_g^+, σ_u^+, and π_g combinations are obtained in the same way. The energy-level diagram for CO_2, as given in Figure 4-13, is then found by solving the relevant secular equations.

The 16 valence electrons occupy the lowest orbitals in the ground state, giving the configuration $(1\sigma_g)^2 (1\sigma_u)^2 (2\sigma_g)^2 (2\sigma_u)^2 (1\pi_u)^4 (1\pi_g)^4$. The molecular wavefunction for the ground state transforms as $^1\Sigma_g^+$.

Low energy excited states occur by promoting an electron from the $1\pi_g$ orbital to the first empty $2\pi_u$ orbital. The orbital configuration $(1\pi_g)^3 (2\pi_u)^1$ is four-fold degenerate. Including spin degeneracy it is sixteen-fold degenerate. We can therefore expect that under the influence of the electron repulsion terms in the Hamiltonian we will get four spin-singlet states and four spin-triplet states. However, not all of these excited states will have a linear conformation.

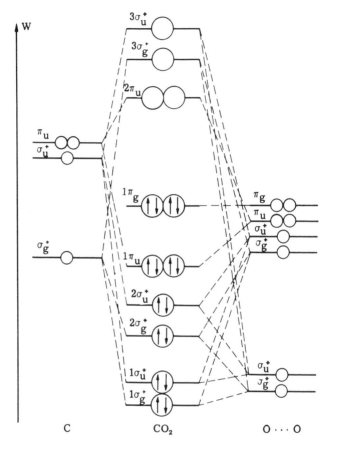

Figure 4-13 Molecular orbital diagram for CO_2.

If we bend the molecule,

$$C$$
$$O \qquad O$$

the relevant symmetry operations are (see Figure 4-14) E, $C_2(y)$,

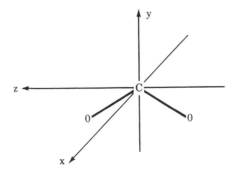

Figure 4-14   Coordinate system for a bent CO_2 molecule.

$\sigma_v(xy)$ and $\sigma_v(yz)$. Orbital degeneracy is no longer possible, as can be seen by applying the symmetry operation to the π orbitals.

	$\hat{E}$	$\hat{C}_2(y)$	$\hat{\sigma}_v(xy)$	$\hat{\sigma}_v(yz)$	symmetry designation
π_x	1	-1	1	-1	b_1
π_y	1	1	1	1	a_1

The π_x and π_y orbitals behave in a different way and are therefore no longer degenerate. Calculations and experiments confirm that a $1\pi_g \rightarrow 2\pi_u(a_1)$ excitation will lead to a bent excited state, whereas the molecule will stay linear if the excited electron occupies the $2\pi_u(b_1)$ orbital.

4-5   FORMALDEHYDE

The H_2CO molecule is planar in its ground state (Figure 4-15). We first construct three strong σ bonds involving the carbon, the two hydrogen, and the oxygen atoms. Since the angles in the plane all are approximately $120°$, we construct three equivalent orbitals in the xy plane that are directed from carbon toward H_1, H_2, and O. For this purpose we hybridize the three carbon atomic orbitals $2s$, $2p_x$, and $2p_y$ to form the trigonal hybrids. The coordinate system is given in Figure 4-15.

On the oxygen we hybridize $2p_x$ and $2s$ (see Figure 4-15) and form the linear hybrids $2p_x - 2s$ and $2p_x + 2s$. The σ electronic structure for formaldehyde is shown in Figure 4-16, with a "lone pair" on O directed away from the carbon.

All together we have, ignoring $C(1s)^2$ and $O(1s)^2$ electrons, $2 \times 1(H) + 4(C) + 6(O) = 12$ valence electrons, of which we have accounted for 8. Still available are $C(p_z)$ and $O(2p_y),(2p_z)$. The π-type orbitals situated on C and O are shown in Figure 4-17. We see that $C(p_z)$ and $O(p_z)$ form π molecular orbitals. The $2p_y$ orbital on oxygen in this approximation is considered to be nonbonding. Assuming the σ bonding orbitals are very stable and therefore that the σ antibonding orbitals have very high energy, we obtain a schematic energy diagram as given in Figure 4-18.

Formaldehyde has the symmetry operations that are given in Table 4-2. Since $O(p_y)$ transforms as b_2 and $O(p_z)$ as b_1, the ground state is $\cdots (b_1^b)^2(b_2)^2; {}^1A_1$.

The lowest electronic excited state occurs on promoting one of the electrons in the b_2 nonbonding orbital to the antibonding orbital (b_1). The electronic configuration is then $\cdots (b_1^b)^2(b_2)^1 (b_1)^1$. Since the two different orbitals (b_2) and $b_1)$ each contain an electron, the excited electronic configuration gives states with S

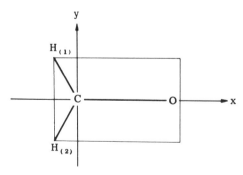

Figure 4-15 The geometry of formaldehyde in the ground state.

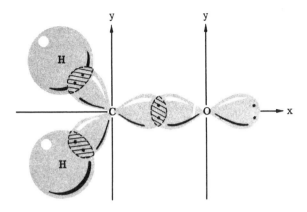

Figure 4-16   The σ electronic structure of formaldehyde.

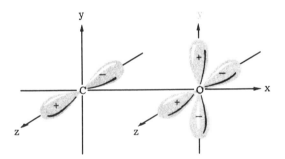

Figure 4-17 π Orbitals in formaldehyde.

Table 4-2   Symmetry operations and characters for planar H_2CO.

	$\hat{E}$	$\hat{C}_2(x)$	$\hat{\sigma}_v(xz)$	$\hat{\sigma}_v(xy)$
A_1	1	1	1	1
A_2	1	1	-1	-1
B_1	1	-1	1	-1
B_2	1	-1	-1	1

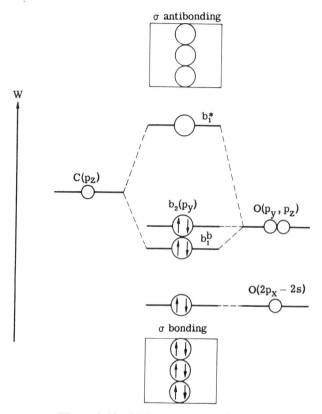

Figure 4-18   Molecular orbitals for H_2CO.

$= 0$ and $S = 1$. Because the product of functions transforming as b_2 and b_1 transforms as a_2 (consult Table 4-2), we obtain $^3A_2$ and $^1A_2$ excited states (Figure 4-19). Hund's rule tells us that the triplet state $^3A_2$ has lower energy than the $^1A_2$ state.

A molecule may be excited from a lower to a higher electronic state by absorption of light (the frequency of the absorbed light is given by Bohr's frequency rule, $\Delta W = h\nu$). Thus each molecule has a characteristic absorption spectrum attributable to its various electronic transitions. The $^1A_1 \rightarrow {}^1A_2$

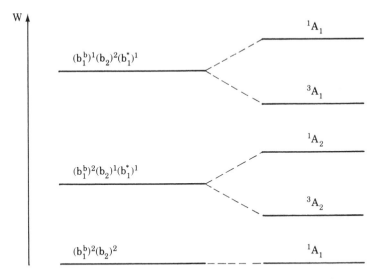

Figure 4-19   The lowest electronic states in the formaldehyde molecule.

transition gives rise to an absorption band in the spectrum of formaldehyde with a maximum around 3000 Å; this spectroscopic feature is a trademark of the carbonyl group, and it is commonly referred to as the $n \rightarrow \pi^*$ transition. Such a transition places a nonbonding oxygen electron in the π antibonding $C - O$ orbital. Thus we expect the transition to be accompanied by an increase in the $C-O$ bond length. This is verified experimentally. In the excited state the $C-O$ distance is ~1.31 Å, but it is only 1.22 Å in the ground state. In addition, in the excited state the oxygen atomic nucleus is no longer in the plane formed by

The out-plane bending is about $20°$ (see Figure 4-20). At higher energy, exciting an electron from b_1^b to b_1^* produces an $^1A_1 \rightarrow {}^1A_1$ transition; this is called the $\pi \rightarrow \pi^*$ transition.

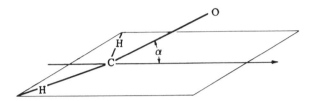

Figure 4-20   The molecular structure of formaldehyde in the excited state $^1A_2$. The barrier for an "umbrella inversion" of the molecule ($\alpha \rightarrow -\alpha$) is ≈ 650 cm^{-1}.

4-6 ETHYLENE

The ethylene molecule is isoelectronic with formaldehyde, and its electronic structural description is similar. In its ground state the C_2H_4 molecule is planar; we construct sp^2 hybrids on each carbon atom and obtain one σ C—C bond and four σ C—H bonds. These bonding orbitals can accommodate 10 electrons, which if we ignore the C($1s$) electrons leave us with 2 electrons.

However, we have not considered the two p_z orbitals, one on each carbon; we can construct a π bond using these two orbitals. The overall bonding scheme is pictured in Figure 4-22. The π molecular orbitals and energies are as follows (p_{z1} and p_{z2} are located on carbon atoms 1 and 2, $\alpha = \int p_{zj}\mathscr{H}p_{zj}d\tau$, $j = 1,2$, and $\beta = \int p_{z1}\mathscr{H}p_{z2}d\tau$):

$$\psi(\pi) = \frac{1}{\sqrt{2 + 2S}} \ (p_{z1} + p_{z2}), \qquad w^b = \frac{\alpha + \beta}{1 + S} ,$$
$$(4\text{-}71)$$

$$\psi(\pi^*) = \frac{1}{\sqrt{2 - 2S}} \ (p_{z1} - p_{z2}), \qquad w^* = \frac{\alpha - \beta}{1 - S} .$$
$$(4\text{-}72)$$

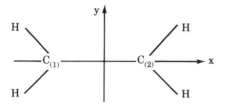

Figure 4-21   Ground state geometry of ethylene; the molecule lies in the x,y plane.

Both α and β are negative quantities. Notice that in this description the designations σ and π are used to describe orbitals that are symmetric and antisymmetric, respectively, upon reflection in the molecular plane.

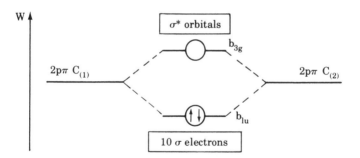

Figure 4-22   Bonding diagram for C_2H_4; the 10 σ bonding electrons occupy 5 low energy orbitals.

Selected symmetry operations or ethylene are $\hat{E}$, $\hat{C}_2(x)$, $\hat{C}_2(y)$, $\hat{C}_2(z)$, and the inversion, $\hat{i}$. The characters are given in Table 4-3.

Table 4-3   Selected symmetry operations and characters for ethylene.

	$\hat{E}$	$\hat{C}_2(z)$	$\hat{C}_2(y)$	$\hat{C}_2(x)$	$\hat{i}$
A_g	1	1	1	1	1
B_{1g}	1	1	-1	-1	1
B_{2g}	1	-1	1	-1	1
B_{3g}	1	-1	-1	1	1
A_u	1	1	1	1	-1
B_{1u}	1	1	-1	-1	-1
B_{2u}	1	-1	1	-1	-1
B_{3u}	1	-1	-1	1	-1

We find, then, for the transformation properties of the orbitals $\psi(\pi)$ and $\psi(\pi^*)$,

	$\hat{E}$	$\hat{C}_2(z)$	$\hat{C}_2(y)$	$\hat{C}_2(x)$	i	
$\psi(\pi)$	1	1	-1	-1	-1	b_{1u}
$\psi(\pi^*)$	1	-1	-1	1	1	b_{3g}

The ground state is ... $(b_{1u})^2$; $^1A_{1g}$.

Excited states: .... $(b_{1u})^1(b_{3g})^1$; $^1B_{2u}$, $^3B_{2u}$,

$\qquad\qquad$ ... $(b_{3g})^2$; $^1A_{1g}$.

The intense band found at about 1650 Å (60,000 cm^{-1}) in the ethylene absorption spectrum is attributable to the transition $^1A_{1g} \rightarrow {}^1B_{2u}$. As π and π^* orbitals are involved, the transition is commonly designated $\pi \rightarrow \pi^*$.

4-7   ALLENE

The carbon chain in allene, C_3H_4, is linear. In the ground state the two H_2C groups lie in planes at right angles to each other (Figure 4-23).

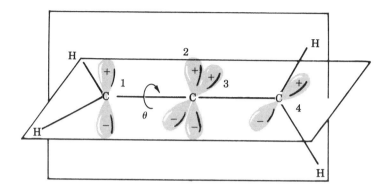

Figure 4-23 Orientation of $p\pi$ orbitals in the ground state conformation of allene.

The terminal carbon atoms have been trigonally hybridized, the carbon atom in the middle linearly hybridized. All in all we have 16 valence electrons; 12 of these are placed in the σ bonds, leaving 4 to be located in π orbitals that can be made up by combining the four $p\pi$ orbitals labeled 1,2,3, and 4 in Figure 4-23.

Our variational wavefunction is then

$$\Psi = c_1\chi_1 + c_2\chi_2 + c_3\chi_3 + c_4\chi_4. \qquad (4\text{-}73)$$

With $\alpha = \int \chi_i \mathcal{H} \chi_i \, d\tau$, $i = 1,2,3,4$, putting all overlap integrals equal to zero, and taking $\beta = \int \chi_1 \mathcal{H} \chi_2 \, d\tau = \int \chi_3 \mathcal{H} \chi_4 \, d\tau$, we have the secular equation

$$\begin{vmatrix} \alpha - w & \beta & 0 & 0 \\ \beta & \alpha - w & 0 & 0 \\ 0 & 0 & \alpha - w & \beta \\ 0 & 0 & \beta & \alpha - w \end{vmatrix} = 0$$

$$(4\text{-}74)$$

with the solutions $w = \alpha - \beta$ and $w = \alpha + \beta$, both solutions being two-fold degenerate. With $\beta < 0$, the ground state energy is $W = 4\alpha + 4\beta$, neglecting electron repulsion terms.

We now investigate the influence of a twist around a C—C bond. Let the left H_2C—group rotate by angle θ around the C—C—C bond axis. Neglecting interactions between the two terminal carbon atoms (because they are far from each other), we get the secular equation for the twisted case

$$\begin{vmatrix} \alpha - w & \beta \cos \theta & \beta \sin \theta & 0 \\ \beta \cos \theta & \alpha - w & 0 & 0 \\ \beta \sin \theta & 0 & \alpha - w & \beta \\ 0 & 0 & \beta & \alpha - w \end{vmatrix} = 0. \quad (4\text{-}75)$$

Expanding, we obtain

$$(\alpha - w)^4 - 2\beta^2 (\alpha - w)^2 + \beta^4 \cos^2 \theta = 0. \quad (4\text{-}76)$$

Solving, we get for the four solutions

$$w = \alpha \pm \beta \sqrt{1 \pm \sin \theta}. \quad (4\text{-}77)$$

With β being a negative number, we have $w_1 < w_2 < w_3 < w_4$, where

$$w_4 = \alpha - \beta \sqrt{1 + \sin \theta}, \quad (4\text{-}78)$$

$$w_3 = \alpha - \beta \sqrt{1 - \sin \theta}, \quad (4\text{-}79)$$

$$w_2 = \alpha + \beta \sqrt{1 - \sin \theta}, \quad (4\text{-}80)$$

$$w_1 = \alpha + \beta \sqrt{1 + \sin \theta}. \quad (4\text{-}81)$$

Neglecting the contributions of the electron repulsion terms to the energy, the sum of the four electron energies for the electronic ground state is given as follows:

$$W = 4\alpha + 2\beta \left[\sqrt{1 + \sin \theta} + \sqrt{1 - \sin \theta} \right]. \quad (4\text{-}82)$$

As can be seen by differentiation, this expression has its lowest value for $\theta = 0$, corresponding to the structure drawn in Figure 4-23.

In the same approximation, the first excited state will have the energy

$$W^{\text{ex}} = 4\alpha + 2\beta \sqrt{1 + \sin \theta}. \qquad (4\text{-}83)$$

Since β is negative, the lowest energy is therefore that for which $\theta = \pi/2$. In other words, the equilibrium conformation in this excited state corresponds to all the atoms in the molecule lying in a plane.

4-8 DIBORANE

Electron diffraction indicates the molecular structure of B_2H_6 as shown in Figure 4-24. H_α, H_β, H_γ, and H_δ lie in the yz plane, whereas B_2H_6 lies in the xz plane.

If we ignore the $B(1s)^2$ electrons, we have $2 \times 3(\text{B}) + 6(\text{H}) = 12$ valence electrons. We assume that the boron $2s$ and $2p$

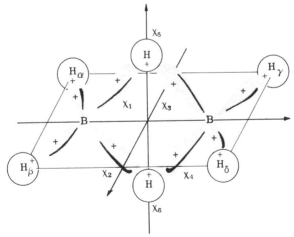

Figure 4-24   The structure of diborane.

orbitals are sp^3 hybridized. Thus 8 electrons can be placed in four σ bonds from the two borons to H_α, H_β, H_γ, and H_δ. This means that we have 4 electrons left to dispose of in the six orbitals $(\chi_1 \cdots \chi_6)$. Clearly, diborane is an "electron-deficient" molecule. There are more valence orbitals than there are electrons.

We construct the following orbitals in an attempt to explain the bonding in the B_2H_2 part of diborane, in which the two hydrogens bridge the two borons:

$$1/2 \ (\chi_1 + \chi_2 + \chi_3 + \chi_4) = \psi_1, \qquad (4\text{-}84)$$

$$1/2 \ (\chi_1 - \chi_2 - \chi_3 + \chi_4) = \psi_2, \qquad (4\text{-}85)$$

$$1/2 \ (\chi_1 + \chi_2 - \chi_3 - \chi_4) = \psi_3, \qquad (4\text{-}86)$$

$$1/2 \ (\chi_1 - \chi_2 + \chi_3 - \chi_4) = \psi_4, \qquad (4\text{-}87)$$

$$\frac{1}{\sqrt{2}} (\chi_5 + \chi_6) = \psi_5, \qquad (4\text{-}88)$$

$$\frac{1}{\sqrt{2}} (\chi_5 - \chi_6) = \psi_6. \qquad (4\text{-}89)$$

All overlap integrals have here been put equal to zero. The symmetry operations of the molecule are the same as those for ethylene, and the characters are given in Table 4-3.

By examining how $\psi_1 \cdots \psi_6$ transform, we find $\psi_1(a_g)$, $\psi_2(b_{2g})$, $\psi_3(b_{1u})$, $\psi_4(b_{3u})$, $\psi_5(a_g)$, and $\psi_6(b_{3u})$. Notice that ψ_1 and ψ_5, and also ψ_4 and ψ_6, can be combined. Thus we obtain the bonding scheme shown in Figure 4-25. The four electrons completely fill the a_g and b_{3u} bonding orbitals and the ground state is ^1A_g.

4-9 OCTAHEDRAL COMPLEXES

Transition metal ions use s, p, and d orbitals in bonding. The addition of d valence orbitals to the s and p orbital set has important consequences, as it means that a larger variety of molecular structures can be accommodated. A structure that is

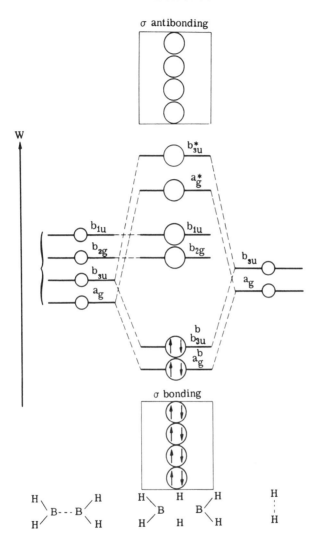

Figure 4-25   Bonding scheme for B_2H_6.

commonly encountered is one in which a central metal atom or ion is surrounded by six groups at the corners of an octahedron. Such an octahedral structure is adopted, for example, by TiF_6^{3-}, $Fe(CN)_6^{4-}$, $V(H_2O)_6^{2+}$, $Co(NH_3)_6^{3+}$, and $Ni(H_2O)_6^{2+}$. Since the metal atom in an octahedral complex is located at a center of high symmetry, the molecular orbitals are conveniently written in the following form:

$$\phi = c_1\chi_M + c_2\Phi_{lig} \qquad (4\text{-}90)$$

where χ_M is the metal orbital, and $\Phi_{lig} = \sum_{n=1}^{m} a_n\chi_n$ is a normalized combination of ligand orbitals (ligands are the groups attached to the metal) that transforms like χ_M. Normalizing the orbital (4-90) to one, we have

$$c_1^2 + c_2^2 + 2c_1c_2G = 1, \qquad (4\text{-}91)$$

$$G = \int \chi_M \ \Phi_{lig} \ d\tau.$$

In Eq. (4-91) G is the overlap integral of the metal orbital with the linear combination of ligand orbitals. This quantity is called the *group overlap*.

We shall now formulate the molecular orbitals for an octahedral complex containing a first-row transition metal ion. The orbitals that will be used in the bonding scheme are the $3d$, $4s$, and $4p$ orbitals of the central atom and the ns and np orbitals of the ligands. A convenient coordinate system for the construction of MO's is shown in Figure 4-26. Some key symmetry operations are given in Figure 4-27, and selected characters for octahedral symmetry are given in Table 4-4.

We examine first the C_3 symmetry operation. By using the coordinate system given in Figure 4-27, we find

$$\hat{C}_3 \begin{pmatrix} x \\ y \\ z \end{pmatrix} = \begin{pmatrix} 0 & 1 & 0 \\ 0 & 0 & 1 \\ 1 & 0 & 0 \end{pmatrix} \begin{pmatrix} x \\ y \\ z \end{pmatrix}. \qquad (4\text{-}92)$$

Table 4-4   Selected symmetry operations and characters for octahedral
(O_h) symmetry.

	$\hat{E}$	$\hat{C}_3$	$\hat{C}_2$	$\hat{C}_4$	$\hat{C}_2'$	$\hat{i}$
A_{1g}	1	1	1	1	1	1
A_{2g}	1	1	1	-1	-1	1
E	2	-1	2	0	0	2
T_{1g}	3	0	-1	1	-1	3
T_{2g}	3	0	-1	-1	1	3
A_{1u}	1	1	1	1	1	-1
A_{2u}	1	1	1	-1	-1	-1
E_u	2	-1	2	0	0	2
T_{1u}	3	0	-1	1	-1	-3
T_{2u}	3	0	-1	-1	1	-3

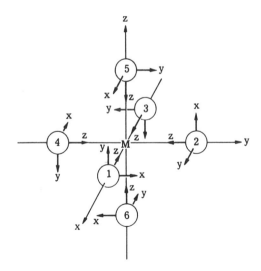

Figure 4-26   Coordinate systems for an octahedral complex.

$$\hat{C}_3 \begin{pmatrix} p_x \\ p_y \\ p_z \end{pmatrix} = \begin{pmatrix} 0 & 1 & 0 \\ 0 & 0 & 1 \\ 1 & 0 & 0 \end{pmatrix} \begin{pmatrix} p_x \\ p_y \\ p_z \end{pmatrix} \quad (4\text{-}93)$$

and for the normalized d orbitals (1-53, 1-54, 1-55, 1-56, 1-57),

$$\hat{C}_3 \begin{pmatrix} d_{xz} \\ d_{yz} \\ d_{xy} \\ d_{z^2} \\ d_{x^2-y^2} \end{pmatrix} = \begin{pmatrix} 0 & 0 & 1 & 0 & 0 \\ 1 & 0 & 0 & 0 & 0 \\ 0 & 1 & 0 & 0 & 0 \\ 0 & 0 & 0 & -\dfrac{1}{2} & \dfrac{\sqrt{3}}{2} \\ 0 & 0 & 0 & -\dfrac{\sqrt{3}}{2} & -\dfrac{1}{2} \end{pmatrix} \begin{pmatrix} d_{xz} \\ d_{yz} \\ d_{xy} \\ d_{z^2} \\ d_{x^2-y^2} \end{pmatrix}$$

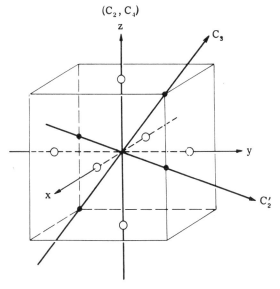

Figure 4-27   Symmetry operations in an octahedron.

It is evident that the three sets $(p_x, p_y, p_z), (d_{z^2}, d_{x^2-y^2})$, and (d_{xz}, d_{yz}, d_{xy}) go into linear combinations of each other under this symmetry operation. By performing the other symmetry operations, outlined in Table 4-4, we find the traces of the transformation matrices as given in Table 4-5. By comparison with the characters (Table 4-4), we see that (s) transforms as a_{1g}, (p_x, p_y, p_z) as t_{1u}, $(d_{x^2-y^2}, d_{z^2})$ as e_g, and (d_{xy}, d_{xz}, d_{yz}) as t_{2g}.

In the next step we find the linear combinations of the ligand ns and np orbitals that can be used for bonding. We shall first consider only σ bonding. Each ligand possesses ns and np_z σ valence orbitals, and we will take a linear combination as given in (4-95):

$$\sigma_{(\text{lig})} = \alpha\chi(s) + \sqrt{1 - \alpha^2}\chi(p_z). \qquad (4\text{-}95)$$

Here α is a mixing parameter, $0 \le \alpha \le 1$.

We now construct six linearly independent molecular orbitals from the six ligand σ functions.

A totally symmetric $\sigma(a_{1g})$ orbital can be written down by inspection:

$$\phi(a_{1g}) = \frac{1}{\sqrt{6}} \ (\sigma_1 + \sigma_2 + \sigma_3 + \sigma_4 + \sigma_5 + \sigma_6). \qquad (4\text{-}96)$$

This ligand combination is normalized if overlap between the six

Table 4-5   Traces of Transformation Matrices.

	$\hat{E}$	$\hat{C}_3$	$\hat{C}_2$	$\hat{C}_4$	$\hat{C}_2$	$\hat{i}$	Type of function
(s)	1	1	1	1	1	1	a_{1g}
(p_x, p_y, p_z)	3	0	−1	1	−1	−3	t_{1u}
$(d_{x^2-y^2}, d_{z^2})$	2	−1	2	0	0	2	e_g
(d_{xy}, d_{xz}, d_{yz})	3	0	−1	−1	1	3	t_{2g}

σ orbitals is neglected. Thus, we have for the complete a_{1g} molecular wavefunction

$$\psi(a_{1g}) = c_1(4s) + c_2 \frac{1}{\sqrt{6}} (\sigma_1 + \sigma_2 + \sigma_3 + \sigma_4 + \sigma_5 + \sigma_6)$$
(4-97)

with

$$c_1^2 + c_2^2 + 2c_1 c_2 G = 1.$$
(4-98)

Taking phase into account, an "extension" of each of the metal p orbitals to the ligands leads to the molecular orbitals.

$$t_{1u}. \begin{cases} \psi(t_{1u}^x) = c_3(4p_x) + c_4 \dfrac{1}{\sqrt{2}} (\sigma_1 - \sigma_3), \\[2mm] \psi(t_{1u}^y) = c_3(4p_y) + c_4 \dfrac{1}{\sqrt{2}} (\sigma_2 - \sigma_4), \\[2mm] \psi(t_{1u}^z) = c_3(4p_z) + c_4 \dfrac{1}{\sqrt{2}} (\sigma_5 - \sigma_6). \end{cases} \quad (4\text{-}99)$$

Notice that the p orbitals and the linear combinations of ligand orbitals transform in the same way under the various symmetry operations.

With $d_{x^2-y^2}$ and d_{z^2} we construct the σ orbitals as follows: The "extension" of the $d_{x^2-y^2}$ orbital to the ligand orbitals yields the $e_g^{x^2-y^2}$ molecular orbitals

$$\psi(e_g^{x^2-y^2}) = c_5(3d_{x^2-y^2}) + c_6 \tfrac{1}{2}(\sigma_1 - \sigma_2 + \sigma_3 - \sigma_4).$$
(4-100)

Let us now rotate this orbital, using the threefold axis (see Figure 4-27):

$$\hat{C}_3\psi(e_g^{x^2-y^2}) = c_5\left(-\tfrac{1}{2}d_{x^2-y^2} - \frac{\sqrt{3}}{2}\,d_{z^2}\right)$$

$$+ c_6\,\tfrac{1}{2}(\sigma_2 - \sigma_5 + \sigma_4 - \sigma_6). \qquad (4\text{-}101)$$

The right-hand side of Eq. (4-101) can be split into two componenents

$$\hat{C}_3\psi(e_g^{x^2-y^2}) = -\frac{1}{2}\left[c_5 d_{x^2-y^2} + c_6\,\tfrac{1}{2}(\sigma_1 - \sigma_2 + \sigma_3 - \sigma_4)\right]$$

$$-\frac{\sqrt{3}}{2}\left[c_5 d_{z^2} + c_6\,\frac{1}{2\sqrt{3}}\,(2\sigma_5 + 2\sigma_6 - \sigma_1 - \sigma_2 - \sigma_3 - \sigma_4)\right].$$

$$(4\text{-}102)$$

We therefore use the two linear combinations in the square brackets of (4-102) as the σ molecular orbitals transforming as e_g. Note that we have now used up all the σ valence orbitals; this is verified by summing all the σ_1 parts:

$$\sigma_1\colon \tfrac{1}{6}(s) + \tfrac{1}{2}(p_x) + \tfrac{1}{4}(d_{x^2-y^2}) + \tfrac{1}{12}(d_{z^2}) = 1.$$

As usual, the variational coefficients, $c_1 \cdots c_6$, are found by minimizing the energy. There is one secular equation to solve for each type of orbital. Here the problem reduces to the solution of three 2×2 determinants for the e_g, a_{1g}, and t_{1u} orbitals, respectively. The molecular orbitals that result are shown in Figure 4-28. Note that in a σ-bonding MO scheme the metal t_{2g} orbitals are strictly nonbonding, and therefore will be lower in energy than the antibonding e_g level. There are $12 + n$ electrons to place in the molecular orbitals (2 from each ligand and n d-electrons from the central atom). As a simple example, we find

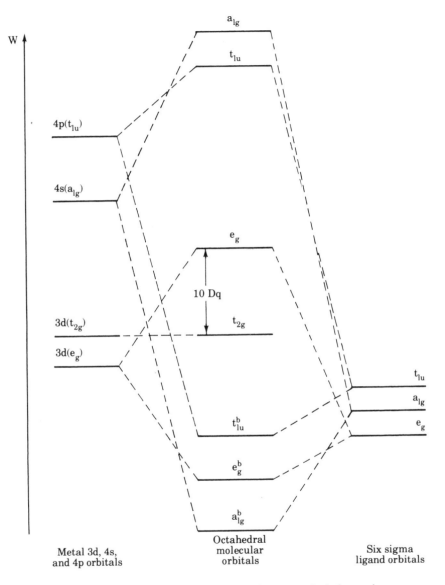

Figure 4-28   Molecular orbital diagram for an octahedral complex; π bonding is neglected.

that the ground state of TiF_6^{3-} is

$$(a_{1g}^b)^2 (t_{1u}^b)^6 (e_g^b)^4 (t_{2g})^1; \, {}^2T_{2g}$$

since Ti^{3+} contributes one electron.

In the formulation described above the t_{2g} orbital is a pure metal orbital. A low-lying excited state in TiF_6^{3-} occurs upon excitation of an electron from t_{2g} to e_g, a ${}^2T_{2g} \rightarrow {}^2E_g$ transition. The energy separation between the e_g and t_{2g} orbitals is known as 10 Dq. The value of 10 Dq in TiF_6^{3-} is obtained from the weak absorption band system with a maximum at about 17,000 cm^{-1}, which is due to the transition ${}^2T_{2g} \rightarrow {}^2E_g$. Thus, $h\nu = 10$ Dq $\approx$ 17,000 cm^{-1} for TiF_6^{3-}. The value of 10 Dq in all cases of interest is obtained from experiment. It is found that Dq $\approx$ 2000 cm^{-1} for the hexaquo complexes of the tripositive metal ions of the first transition series, and for the corresponding dipositive complexes, Dq $\approx$ 1000 cm^{-1}. The 10 Dq values for complexes depend strongly on the nature of the ligand, for example,

$$10 \text{ Dq} \, (CN^-) > 10 \text{ Dq} \, (NH_3)$$
$$> 10 \text{ Dq} \, (H_2O) > 10 \text{ Dq} \, (F^-).$$

Higher electronic excited states arise on excitation of an electron from a bonding MO into t_{2g} or e_g. Since the bonding MO's are mainly ligand orbitals and t_{2g} and e_g are mainly metal orbitals, this type of transition is known as ligand-to-metal charge transfer (abbreviated $L \rightarrow M$). The first such band occurs higher than 50,000 cm^{-1} in the spectrum of TiF_6^{3-}.

Formulation of the π molecular orbitals in an octahedral complex is straightforward. The modifications produced in the MO energy level scheme are important for consideration of charge transfer transitions; further, if we want to perform explicit calculations, it is of prime importance to include π bonding. Both σ and π metal and normalized ligand combinations for an octahedral complex are summarized in Table 4-6.

The general energy level diagram for an octahedral complex with inclusion of π bonding from the ligands to the metal is shown

Table 4-6 Metal and Ligand Orbitals for the Molecular Orbitals of an Octahedral Complex.

Representation	Metal orbital	Ligand orbitals	
		σ	π
a_g	$4s$	$\frac{1}{\sqrt{6}}(\sigma_1 + \sigma_2 + \sigma_3 + \sigma_4 + \sigma_5 + \sigma_6)$	
e_g	$3d_{x^2-y^2}$	$\frac{1}{2}(\sigma_1 - \sigma_2 + \sigma_3 - \sigma_4)$	
	$3d_{z^2}$	$\frac{1}{2\sqrt{3}}(2\sigma_5 + 2\sigma_6 - \sigma_1 - \sigma_2 - \sigma_3 - \sigma_4)$	
t_{1u}	$4p_x$	$\frac{1}{\sqrt{2}}(\sigma_1 - \sigma_3)$	$\frac{1}{2}(y_2 + x_5 - x_4 - y_6)$
	$4p_y$	$\frac{1}{\sqrt{2}}(\sigma_2 - \sigma_4)$	$\frac{1}{2}(x_1 + y_5 - y_3 - x_6)$

$4p_z$	$\dfrac{1}{\sqrt{2}}(\sigma_5 - \sigma_6)$	$\frac{1}{2}(y_1 + x_2 - x_3 - y_4)$
t_{2g}	$3d_{xz}$	$\frac{1}{2}(y_1 + x_5 + x_3 + y_6)$
	$3d_{yz}$	$\frac{1}{2}(x_2 + y_5 + y_4 + x_6)$
	$3d_{xy}$	$\frac{1}{2}(x_1 + y_2 + y_3 + x_4)$
t_{1g}		$\frac{1}{2}(y_1 - x_5 + x_3 - y_6)$
		$\frac{1}{2}(x_2 - y_5 + y_4 + x_6)$
		$\frac{1}{2}(x_1 - y_2 + y_3 - x_4)$
t_{2u}		$\frac{1}{2}(y_2 - x_5 - x_4 + y_6)$
		$\frac{1}{2}(x_1 - y_5 - y_3 + x_6)$
		$\frac{1}{2}(y_1 - x_2 - x_3 + y_4)$

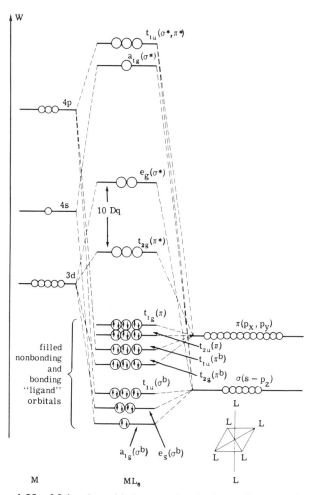

Figure 4-29   Molecular orbital energy level scheme for an octahedral complex. All the bonding orbitals are filled with ligand electrons.

in Figure 4-29. The "metal" electrons are now accommodated in $t_{2g}(\pi^*)$ and $e_g(\sigma^*)$. In this scheme it is recognized that the unpaired electron in TiF_6^{3-} is in a π-antibonding MO ($^2T_{2g}$ ($t_{2g}(\pi^*)$)) ground state. However, it is common to refer to the $t_{2g}(\pi^*)$ and $e_g(\sigma^*)$ levels simply as t_{2g} and e_g.

The ground state for CrF_6^{3-} is $^4A_{2g}(t_{2g})^3$, and for NiF_6^{4-} it is $^3A_{2g}(t_{2g})^6(e_g)^2$. The ground states adopted by octahedral complexes of metal ions with d^4, d^5, d^6, and d^7 configurations depend on two factors: (1) the magnitude of 10 Dq and (2) the energy required to pair two electrons in the t_{2g} level. Thus, if 10 Dq is smaller than the t_{2g} pairing energy, the extra electrons will occupy e_g in preference to pairing in the t_{2g} level.[1] Such complexes are said to have *high-spin* ground states. Examples are $Fe(H_2O)_6^{2+}$ $^5T_{2g}(t_{2g})^4(r_g)^2$ and $Fe(H_2O)_6^{3+}$ $^6A_{1g}(t_{2g})^3(e_g)^2$. If 10 Dq is larger than the t_{2g} pairing energy, however, then the fourth, fifth, and sixth electrons will prefer to occupy t_{2g}, giving a *low-spin* ground state. Examples are $Fe(CN)_6^{4-}$ $^1A_{1g}(t_{2g})^6$ and $Fe(CN)_6^{3-}$ $^2T_{2g}(t_{2g})^5$. Low-lying electronic absorption bands in these and other d^n octahedral complexes are due to excitation of electrons from t_{2g} to the e_g level. The detailed description of these excited states for $n = 2\text{-}8$ requires explicit consideration of e^2/r_{ij} terms and will not be considered here.

The origin of the energy separation between the e_g and t_{2g} levels can be understood in the following way. If the bonding orbital is approximated as a pure ligand function, Φ_{lig}, the orthogonalization inherent in a molecular orbital calculation introduces a nodal plane in the antibonding "metal" wavefunction, $\Phi_M = \chi_d - G(d, lig)\Phi_{lig}$. Since σ orbitals overlap to a greater extent than do π orbitals, it follows that $G(e_g) > G(t_{2g})$. The metal electrons are by orthogonality prohibited from entering the space occupied by the ligand electrons; therefore, e_g electrons are more confined than are t_{2g} electrons, and their energies are correspondingly higher. In other words, the octahedral ligand field splitting is a manifestation of the Pauli principle.

---

[1] See the discussion on p. 56 of Chap. 3 for the principles involved.

4-10 POTENTIAL SURFACES

In the calculations of the electronic energies of a molecule or ion we have seen that the positions of the nuclei appear as parameters. Thus we can look at the molecular electronic energies as functions of the nuclear coordinates, and we refer to such functions as potential surfaces.

The positions of N nuclei in space are determined by $3N$ cartesian coordinates. Three of these coordinates can be used to specify the center of mass of the molecule and three coordinates to specify the orientation in space. (For a linear molecule we need only two coordinates to account for the orientation.) For all but linear molecules the potential surfaces are therefore dependent upon $3N - 6$ nuclear coordinates.

If we want to consider a chemical reaction that takes place involving three nuclei

$$AB + C \rightarrow A + BC$$

we will need three nuclear coordinates in the expressions for the electronic energies. These can be taken as three internuclear distances, R_{AB}, R_{BC}, R_{CA}. But we could also have chosen two distances and one angle, R_{AB}, R_{BC}, and angle ABC. Indeed, one distance and two angles could also be used but is less convenient in practice.

By calculating the electronic energy of the ground state of the three nuclei as a function of say R_{AB}, R_{BC}, and angle ABC, we will find a potential energy surface that is made up of two long narrow valleys. These represent the stable molecules AB and BC. They are connected by a "pass" leading through a region of high energy. In order "to go over the pass" the system must acquire *activation energy*. Classically, we have of course that a system with less energy than the "height" of the pass cannot go over it. However, quantum mechanically the system may "tunnel" through. This last process is particularly important in reactions involving transfer of electrons.

We can understand the order of magnitude for the activation energy by the following example. Consider the reaction

$$H_2 + D_2 \rightarrow 2HD. \qquad (4\text{-}103)$$

Assume that the reaction proceeds in such a way that all four nuclei lie in a plane

(1) H D (3)	(1) H — D (3)	(1) H — D (3)
| + | →	| | →	+
(2) H D (4)	(2) D — H (4)	(2) D — H (4)

$$\qquad I \qquad\qquad\qquad II \qquad\qquad\qquad III$$

We can then write down the four linear combinations of $(1s)$ orbitals, one orbital centered on each of the four nuclei, which must be involved in the transition from I to III

$$\psi_1 = (1s)_1 + (1s)_2 + (1s)_3 + (1s)_4, \qquad (4\text{-}104)$$

$$\psi_2 = (1s)_1 + (1s)_2 - (1s)_3 - (1s)_4, \qquad (4\text{-}105)$$

$$\psi_3 = (1s)_1 - (1s)_2 + (1s)_3 - (1s)_4, \qquad (4\text{-}106)$$

$$\psi_4 = (1s)_1 - (1s)_2 - (1s)_3 + (1s)_4. \qquad (4\text{-}107)$$

In situation I with bonding between H—H and D—D the orbital energies will be as pictured in Figure 4-30. In the square planar configuration II, ψ_2 and ψ_3 are degenerate, since their nodal properties are identical. Finally, for configuration III the orbital picture is altered to suit the requirements of H—D bonding. It is obvious that the electronic configuration I of $(\psi_1)^2(\psi_2)^2$ has lower energy than $(\psi_1)^2(\psi_2,\psi_3)^2$. Calculations show indeed that the energy of the molecular structure II is about 6.5 eV higher than the energy of $2H_2$. For comparison, we remember that the dissociation energy of H_2 is 4.7 eV. Therefore, it is not very likely that a square planar geometry is the "transition state" nuclear conformation; a path leading through a trapezoidal structure seems the most likely.

We have previously noted that any molecule with a fixed nuclear arrangement will have certain symmetry operations that leave the molecular Hamiltonian unaltered. However, in some

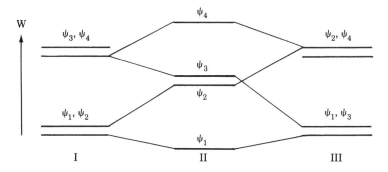

Figure 4-30   Orbitals for the reaction $H_2 + D_2 \rightarrow 2HD$.

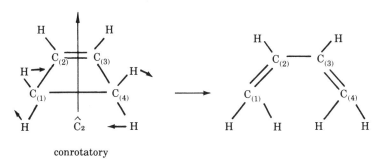

conrotatory

Figure 4-31   Conrotatory opening of cyclobutene.

cases a molecule may also change its shape in such a way that it maintains one or more symmetry operations during the whole rearrangement of nuclei. A classification of the orbitals and states under the symmetry operations that "survive" can therefore be done; the symmetry designations retain their validity and are "good quantum numbers" during the chemical reaction.

The classic example here is the ring opening of cyclobutene to *cis*-butadiene. This process can be envisaged to go in two ways. Either we have a *conrotatory mode* in which the two CH_2 groups rotate the same way in phase, or we have a

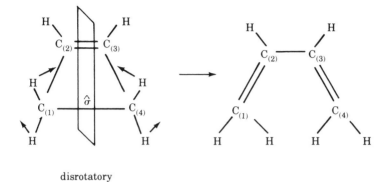

disrotatory

Figure 4-32 Disrotatory opening of cyclobutene.

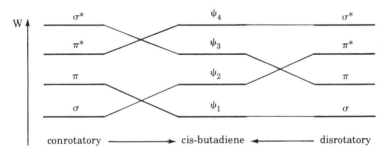

Figure 4-33 Correlation diagram: cyclobutene to cis-butadiene.

disrotatory mode in which the two CH_2 groups rotate "against" each other but still in phase. In the first case the C_2 axis constitutes a symmetry axis, and in the second case the symmetry plane σ_v retains its validity during the whole conversion.

The orbitals that are involved are seen to make the π bond between carbon atoms 2 and 3 and the σ bond between carbon atoms 1 and 4. We classify them according to whether they are symmetric ($+$) or antisymmetric ($-$) with respect to the symmetry operation $\hat{C}_2$ or $\hat{\sigma}_v$. We find easily

cyclobutene	$\hat{C}_2$	$\hat{\sigma}_v$	cis-butadiene	$\hat{C}_2$	$\hat{\sigma}_v$
$\sigma = \sigma_1 + \sigma_4$	+	+	$\psi_1 = \pi_1 + \pi_2 + \pi_3 + \pi_4$	−	+
$\sigma^* = \sigma_1 - \sigma_4$	−	−	$\psi_2 = \pi_1 + \pi_2 - \pi_3 \, p \, \pi_4$	+	−
$\pi = \pi_2 + \pi_3$	−	+	$\psi_3 = \pi_1 - \pi_2 - \pi_3 + \pi_4$	−	+
$\pi^* = \pi_2 - \pi_3$	+	−	$\psi_4 = \pi_1 - \pi_2 + \pi_3 - \pi_4$	+	−

The energies of the orbitals are $w(\sigma^*) > w(\pi^*) > w(\pi) > w(\sigma)$ and $w(\psi_4) > w(\psi_3) > w(\psi_2) > w(\psi_1)$. The correlation diagram for this case is shown in Figure 4-33. For a conrotatory conversion, the gound state of cyclobutene $^1A(\sigma)^2(\pi)^2$ correlates with the lowest orbital configuration of cis-butadiene, $^1A(\psi_1)^2(\psi_2)^2$. In a disrotatory process, we find that the ground state of cyclobutene, $^1A(\sigma)^2(\pi)^2$, correlates with the $^1A(\psi_1)^2(\psi_3)^2$ cis-butadiene. Thus both the forward and the reverse reactions are predicted to be conrotatory in the electronic ground state. Experiments confirm this prediction.

4-11 ELECTRIC DIPOLES

We pointed out in section 1-1 that a knowledge of the molecular wavefunctions allows one to calculate all the physical properties of the molecule. As an example, we shall calculate the permanent electric dipole of a molecule in its ground state.

The dipole moment of a molecule is a measure of the separation of the centroids of its positive and negative charges. With the positive charges centered on the nuclei, and the negative charges on the electrons, the dipole moment $(\vec{\mu})$ for an electrically neutral system consisting of three nuclei (A, B and C) and three electrons ($j = 1,2,3$) placed in a coordinate system (Figure 4-34) is as follows

$$\vec{\mu} = |e| \, \{\vec{\rho}_A + \vec{\rho}_B + \vec{\rho}_C - \langle\vec{\rho}_1\rangle - \langle\vec{\rho}_2\rangle - \langle\vec{\rho}_3\rangle\}$$

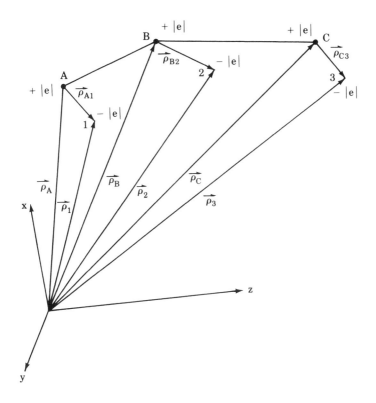

Figure 4-34 A laboratory-fixed coordinate system for the calculation of dipole moments. The nuclei are designated by A, B, C; the electrons by 1, 2, 3.

where by $\langle \vec{\rho}_j \rangle$, $j = 1,2,3$ we have indicated the averaging of $\vec{\rho}_j$ over the ground state wavefunction $\Psi_0(1,2,3)$;

$$\langle \vec{\rho}_j \rangle = \int \Psi_0 \vec{\rho}_j \Psi_0 \, d\tau_1 \, d\tau_2 \, d\tau_3. \qquad (4\text{-}108)$$

With (see Figure 4-34) $\vec{\rho}_A = \vec{\rho}_1 - \vec{\rho}_{A1}$, $\vec{\rho}_B = \vec{\rho}_2 - \vec{\rho}_{B2}$, and $\vec{\rho}_C = \vec{\rho}_3 - \vec{\rho}_{C3}$, we obtain

$$\mu = - \,|e\,|\{\, \langle\vec{\rho}_{A1}\rangle + \langle\vec{\rho}_{B2}\rangle + \langle\vec{\rho}_{C3}\rangle \,\}. \quad (4\text{-}109)$$

Quite generally, we notice therefore that only for an electrically neutral system can we calculate a value of the electric dipole that does not depend upon the laboratory-fixed coordinate system, but only upon the nuclear coordinates of the molecule.

A molecular coordinate system can always be chosen such that the centroid of positive nuclear charges fixes the (0,0,0) coordinates. We have, then, $\vec{\rho}_j = \mathbf{i}x_j + \mathbf{j}y_j + \mathbf{k}z_j$, and for a molecule possessing a center of symmetry the inversion operation can be applied:

$$\hat{\imath} \int \Psi_0 \sum_j \vec{\rho}_j \Psi_0 \, d\tau_1 \cdots d\tau_j = - \int \Psi_0 \sum_j \vec{\rho}_j \Psi_0 \, d\tau_1 \cdots d\tau_j$$

$$(4\text{-}110)$$

It follows that the centroid of negative charges also falls at (0,0,0). Therefore, for a molecule with a center of symmetry, there is no separation of the positive and negative centroids; such a molecule can have no permanent electric dipole.

An example of a molecule that possesses a permanent electric dipole is LiH. Let us assume that the charge distribution of $Li^{3+}(1s)^2$ can be approximated by Li^+ and that the bonding orbital is given by

$$\psi_\sigma = \frac{1}{\sqrt{1 + \lambda^2 + 2\lambda S}} \, (\chi_{1s}^H + \lambda\chi_{2p\sigma}^{Li}) \quad (4\text{-}111)$$

with $S = \int \chi_{1s}^H \chi_{2p\sigma}^{Li} \, d\tau$ and where λ measures the polarity of the orbital.

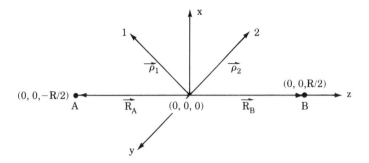

Figure 4-35 Coordinate system for calculating the dipole of the LiH
molecule.

$$\vec{\mu} = |e| \, \{\vec{R}_A + \vec{R}_B - \int \Psi_0 (\vec{\rho}_1 + \vec{\rho}_2) \Psi_0 \, d\tau_1 \, d\tau_2 \}$$

(4-112)

with $\vec{R}_A + \vec{R}_B = 0$, we find

$$\mu = -2|e| \int \Psi_\sigma(1)\vec{\rho}_1 \Psi_\sigma(1) \, d\tau_1$$

$$= \frac{-|e| \, \{R(\lambda^2 - 1) + 4\lambda \int \chi^H_{1s} z \, \chi^{Li}_{2p\sigma} \, d\tau \}}{1 + \lambda^2 + 2\lambda S} \, .$$

The value of λ can only be found by doing a full calculation
of the energy of the electronic ground state. This is quite
laborious. Therefore, one usually turns the procedure around and
uses the measured value of μ to estimate λ.

The observed dipole moment of LiH is 5.9×10^{-18} e.s.u. or 5.9 Debye. The overlap integral S is calculated to be about 0.5 and $\int \chi_{1s}^{H} z \chi_{2p}^{Li} \, d\tau \simeq -0.5$ Å. Using the measured value of the dipole moment, the bonding molecular orbital is then $\Psi_\sigma = 0.72 \chi_{1s}^{H} + 0.42 \chi_{2p\sigma}^{Li}$. As expected, hydrogen is the more electronegative atom in LiH, and the charge distribution is therefore $Li^{\delta+} H^{\delta-}$.

Chapter 4

Suggested Reading

C. A. Coulson, Valence, 2nd ed., Oxford University Press, New York, 1961.
The classic book in the field.

S. P. McGlynn, L. G. Vanquickenborne, M. Kinoshita and D. G. Carroll, Introduction to Applied Quantum Chemistry, Holt, Reinhard, and Winston, New York, 1972.
Instructive survey.

F. A. Cotton, Chemical Applications of Group Theory, 2nd ed., Wiley-Interscience, New York, 1971.

C. J. Ballhausen and H. B. Gray, Coordination Chemistry, Vol. I, ed. A. E. Martell, Van Nostrand-Reinhold, New York, 1971.
Electronic structures of inorganic complexes.

G. E. Kimball, *J. Chem. Phys., 8*, (1940) 188.
Directed valence.

C. A. Coulson, *Trans. Faraday Soc., 33* (1937) 1479; C. A. Coulson and I. Fischer, *Philos. Mag., 40*, (1949) 386.
Treatment of the hydrogen molecule.

A. D. Walsh, *J. Chem. Soc., 1953*, 2260, 2266.
The first two papers in a series on the relationship between molecular orbital theory and the shapes of polyatomic molecules.

D. S. McClure, *Solid State Physics, 9* (1959), 399.
A comprehensive review of the *d-d* spectra of octahedral complexes.

J. N. Murrell, *Structure and Bonding*, *32*, (1977) 93; H. C. Longuet-Higgins and E. W. Abrahamson, *J. Am. Chem. Soc.*, *87* (1965), 2045.
Good treatments of potential surfaces.
C. J. Ballhausen, *J. Chem. Ed.*, *56*, (1979) 215, 294, 357.
Historical perspective.

Problems

1. The ground state of C_2 is not known for certain. The $^3\Pi_u$ and $^1\Sigma_g^+$ states are believed to have approximately the same energy. Which orbital configurations are responsible for these states, and under what conditions would they have nearly the same energy?

2. What are the term designations for the ground states of the following molecules: S_2, B_2, Cl_2^+, O_2^+, BF, BN, BO, CH, and NH?

3. The lowest electronic states in O_2 are $^3\Sigma_g^-$ (0 cm^{-1}), $^1\Delta_g$ (7900 cm^{-1}), and $^1\Sigma_g^+$ (13,200 cm^{-1}). These states all come from the configuration $(\pi_g)^2$, where π_g^x and π_g^y are the two degenerate π antibonding orbitals. The wavefunctions for the ground state are given in the text; the excited state wavefunctions are

$$\Psi(^1\Delta_g) = \begin{cases} \dfrac{1}{\sqrt{2}} \ [\,|\,\overset{+}{\pi}{}_g^x\ \overset{-}{\pi}{}_g^x\,| + |\,\overset{+}{\pi}{}_g^y\ \overset{-}{\pi}{}_g^y\,|\,], \\[4mm] \dfrac{1}{\sqrt{2}} \ [\,|\,\overset{+}{\pi}{}_g^x\ \overset{-}{\pi}{}_g^y\,| - |\,\overset{-}{\pi}{}_g^x\ \overset{+}{\pi}{}_g^y\,|\,], \end{cases}$$

and

$$\Psi(^1\Sigma_g^+) = \dfrac{1}{\sqrt{2}} \ [\,|\,\overset{+}{\pi}{}_g^x\ \overset{-}{\pi}{}_g^x\,| - |\,\overset{+}{\pi}{}_g^y\ \overset{-}{\pi}{}_g^y\,|\,].$$

Suppose now that the O_2 molecules are placed in "unsymmetric" surroundings, such that π_g^x and π_g^y are no longer degenerate; we say that the system has been perturbed. Calling the one electron "perturbation" operation $\hat{V}$, we have

$$\int \pi_g^x \, \hat{V} \, \pi_g^x \, d\tau = -\frac{v}{2}, \qquad \int \pi_g^y \, \hat{V} \pi_g^y \, d\tau = \frac{v}{2},$$

and

$$\int \pi_g^x \, \hat{V} \pi_g^y \, d\tau = 0.$$

Write down a linear combination of the unperturbed states and calculate the diagonal and nondiagonal matrix elements using $\hat{V}$ as the operator. Find the energies of the perturbed O_2 states as a function of v, and sketch the state energies as a function of v.

The perturbation can change the ground state of the system. Find the value of v where this happens. What happens physically to the system at this value of v?

The perturbed system shows a band at $16,000 \ cm^{-1}$. What is the value of v? Do you expect to see other bands? [See J. S. Griffith, *J. Chem. Phys.*, *40*, (1964) 2899.]

4. When a carbon arc light burns in air, the radical CN is formed. The spectrum of the arc shows two strong absorption bands at about $9000 \ cm^{-1}$ and $26,000 \ cm^{-1}$.

    a. What is the term designation of the ground state?

    b. Make tentative assignments for the two transitions.

5. What is the term designation of the ground state of N_3, assuming the molecule is linear? What is the term designation for a bent N_3 molecule?

6. The molecule HNO is diamagnetic in its ground state. We note that $H + N$ equals O. What shape does the molecule have? What would be the first electronic transition? If you apply similar ideas to HCN, what do you obtain?

7. Formulate the bonding in NH_3 in terms of delocalized molecular orbitals. The molecule is trigonal pyramidal. Compare the general molecule orbital description with a localized "tetrahedral" model for NH_3. Discuss the values of the following bond angles: H—N—H, 107°; H—P—H (in PH$_3$), 94°; and F—N—F (in NF$_3$), 103°.

8. Boron trifluoride has a trigonal-planar structure. Formulate the bonding in terms of molecular orbitals. In addition, construct wavefunctions for three equivalent sp^2 hybrid orbitals, using the $2p_x$, $2p_y$, and $2s$ boron valence orbitals, which may be used to form three localized bonds with the three fluorines. Compare and contrast the molecular orbital and the hybrid orbital descriptions.

9. The SF$_6$ molecule has an octahedral structure. Construct the various σ and π bonding orbitals. What is the ground state of the molecule?

10. Construct wavefunctions for six equivalent octahedral (d^2sp^3) hybrid orbitals, using $d_{x^2-y^2}$, d_{z^2}, s, p_x, p_y, and p_z valence orbitals.

An octahedral complex ion in $(MA_6)^3$ has excited electronic states corresponding to the excitation of an electron from an orbital on the M^{3+} ion into an empty orbital on one of the ligands A, say the jth one (charge transfer excitation). Let us designate the zero-order wavefunction representing an electron from M excited to an orbital on ligand j as ϕ_j. Because the electron may "tunnel" from one ligand to another, there is an effective interaction between any two excited states ϕ_i, ϕ_j. The matrix elements are as follows:

$$\int \phi_j \hat{V} \phi_j \, d\tau = 0,$$

$$\int \phi_j \hat{V} \phi_i \, d\tau = v_1 \quad \text{if } i,j \text{ are adjacent vertices,}$$

$$\int \phi_j \hat{V} \phi_i \, d\tau = v_2 \quad \text{if } i,j \text{ are opposite vertices.}$$

In zero-order all the states ϕ_j have excitation energy w above the ground state. Find the correct linear combinations of the ϕ_j, $j = 1,\ldots,6$, which are wavefunctions for the excited states, and calculate their energies.

12. The equilibrium internuclear distance (R_{eq}) in LiBr is 4.10 a_0, and the "covalent" potential curve for $R > 4.10$ a_0 can be taken to be constant. The ionic potential curve in the same interval can be approximated by $V(R) = -e^2/R$. Lithium has an ionization energy of 5.390 eV, and atomic bromine an electron affinity of 3.363 eV, that is, the energy released for addition of an electron to Br. By heating LiBr, what will the dissociation products be? At what distance, R_{eq}, will the "ionic" and "covalent" potential curves cross each other? Would you expect R_{eq} for RbBr to be larger or smaller than R_{eq} for LiBr? [R. S. Berry, *J. Chem. Phys., 27* (1957), 1288.]

INDEX